HIGH PERFORMANCE MEMORY TESTING:

Design Principles, Fault Modeling and Self-Test

FRONTIERS IN ELECTRONIC TESTING

Consulting Editor
Vishwani D. Agrawal

Books in the series:

SOC (System-on-a-Chip) Testing for Plug and Play Test Automation
K. Chakrabarty
ISBN: 1-4020-7205-8

Test Resource Partitioning for System-on-a-Chip
K. Chakrabarty, Iyengar & Chandra
ISBN: 1-4020-7119-1

A Designers' Guide to Built-in Self-Test
C. Stroud
ISBN: 1-4020-7050-0

Boundary-Scan Interconnect Diagnosis
J. de Sousa, P.Cheung
ISBN: 0-7923-7314-6

Essentials of Electronic Testing for Digital, Memory, and Mixed Signal VLSI Circuits
M.L. Bushnell, V.D. Agrawal
ISBN: 0-7923-7991-8

Analog and Mixed-Signal Boundary-Scan: A Guide to the IEEE 1149.4 Test Standard
A. Osseiran
ISBN: 0-7923-8686-8

Design for At-Speed Test, Diagnosis and Measurement
B. Nadeau-Dosti
ISBN: 0-79-8669-8

Delay Fault Testing for VLSI Circuits
A. Krstic, K-T. Cheng
ISBN: 0-7923-8295-1

Research Perspectives and Case Studies in System Test and Diagnosis
J.W. Sheppard, W.R. Simpson
ISBN: 0-7923-8263-3

Formal Equivalence Checking and Design Debugging
S.-Y. Huang, K.-T. Cheng
ISBN: 0-7923-8184-X

Defect Oriented Testing for CMOS Analog and Digital Circuits
M. Sachdev
ISBN: 0-7923-8083-5

Reasoning in Boolean Networks: Logic Synthesis and Verification Using Testing Techniques
W. Kunz, D. Stoffel
ISBN: 0-7923-9921-8

Introduction to I_{DDQ} Testing
S. Chakravarty, P.J. Thadikaran
ISBN: 0-7923-9945-5

Multi-Chip Module Test Strategies
Y. Zorian
ISBN: 0-7923-9920-X

Testing and Testable Design of High-Density Random-Access Memories
P. Mazumder, K. Chakraborty
ISBN: 0-7923-9782-7

From Contamination to Defects, Faults and Yield Loss
J.B. Khare, W. Maly
ISBN: 0-7923-9714-2

HIGH PERFORMANCE MEMORY TESTING:

Design Principles, Fault Modeling and Self-Test

R. Dean Adams
IBM

Kluwer Academic Publishers

Boston / Dordrecht / London

Distributors for North, Central and South America:
Kluwer Academic Publishers
101 Philip Drive
Assinippi Park
Norwell, Massachusetts 02061 USA
Telephone (781) 871-6600
Fax (781) 681-9045
E-Mail: kluwer@wkap.com

Distributors for all other countries:
Kluwer Academic Publishers Group
Post Office Box 322
3300 AH Dordrecht, THE NETHERLANDS
Telephone 31 786 576 000
Fax 31 786 576 254
E-Mail: services@wkap.nl

Electronic Services < http://www.wkap.nl>

Library of Congress Cataloging-in-Publication Data

R. Dean Adams / HIGH PERFORMANCE MEMORY TESTING:
Design Principles, Fault Modeling and Self-Test
ISBN: 1-4020-7255-4

Printed on acid-free paper.

Printed in the United States of America.

For my wife, my son, my daughter, my mother,
and my late father who was an electrical engineer long before me.

Preface

The idea for this book first lodged in my mind approximately five years ago. Having worked on the design of advanced memory built-in self-test since 1988, I saw a need in the industry and a need in the literature. Certain fallacies had grown up and the same mistakes were frequently being repeated. Many people "got away with" these mistakes. As the next generation of chips was produced, however, the large number of bits on a chip made the fruit of these mistakes very evident and the chip quality suffered as a result.

Memory test, memory design, and memory self test are each intriguing subjects. Sinking my teeth into a new memory design article in the Journal of Solid State Circuits is a privilege. Sitting through a clear presentation at the International Test Conference on memory testing can provide food for thought about new ways that memories can fail and how they can be tested. Reviewing a customer's complex memory design and generating an efficient self-test scheme is the most enjoyable work I do. Joining my colleagues at the IEEE Memory Technology, Design, and Test Workshop provides some real memory camaraderie. I hope that the reader will gain some insight from this book into the ways that memories work and the ways that memories fail. It is a fascinating area of research and development.

The key message of this book is that we need to understand the memories that we are testing. We cannot adequately test a complex memory without first understanding its design. Comprehending the design and the test of a memory allows the best memory built-in self-test capabilities. These are needed now and will be needed even more in the future.

This book is in many ways the culmination of 20 years experience in the industry. Having worked in memory design, memory reliability development, and most significantly in memory self test, has allowed me to see memories in a different light. In each role, the objective has been to generate robust, defect-free memories. Memory self test has grown from infancy in the mid 1980s, when even its worth was questioned, to become a relied upon contributor to memory quality and satisfied chip customers.

Here's to good memories!

Acknowledgements

I would like to thank all the people who have helped me over the years to understand memories. These people have included engineers who taught me the basic concepts years ago through customers today who constantly explore new ways to make memories do amazing things.

Many people helped me with this text. Carl Harris and Dr. Vishwani Agrawal each helped significantly in roles with Kluwer. Many colleagues reviewed individual chapters including Tony Aipperspach, John Barth, Kerry Bernstein, Geordie Braceras, Rob Busch, Dr. Lou Bushard, Dr. Kanad Chakraborty, Prof. Bruce Cockburn, Prof. Edmond Cooley, John Debrosse, Jeff Dreibelbis, Tom Eckenrode, Rich Henkler, Bill Huott, Gary Koch, Dr. Bernd Koenemann, Chung Lam, Wendy Malloch, Sharon Murray, Dr. Phil Nigh, Mike Ouellette, Harold Pilo, Jeremy Rowland, Phil Shephard, Prof. Ad van de Goor, Brian Vincent, Tim Vonreyn, Dave Wager, Larry Wissel, Steve Zier, and Johanne Adams. Their insights, enthusiasm, and effort are greatly appreciated.

My management at IBM has been quite supportive of this effort, especially Bernd Koenemann and Frank Urban. My fellow Design-For-Test and BIST colleagues, Carl Barnhart, Ron Walther, Gary Koch, and Tom Eckenrode are each appreciated. Prof. Ad van de Goor, for doing so much fundamental memory test research, publishing, and encouraging is recognized.

My family has been quite helpful looking for chapter opening quotes, reading sections, and being very patient.

Mostly, thanks are due to God.

R. Dean Adams
St. George, Vermont
July 2002

Table of Contents

Chapter 1

Opening Pandora's Box

Design & Test of Memories

"*Thanks for the memories ....*" – Bob Hope's theme song

Memories store ones and zeros. This is basic and simple. Semiconductor memories have existed for decades. These memories have been designed, produced, tested, and utilized by customers all over the world with success. What could possibly be "new and improved" with respect to the design and test of memories? What could possibly be said or even summarized which hasn't been so stated many times before? Much it turns out.

This book is about the self test of memories, the test of memories, and the design of memories. To understand the self-test concept one must first understand memory testing. To properly understand memory testing, though, one must first understand memory design. This understanding is key to comprehending the ways that memories can fail. The testing and operation of memories is radically different from logic. The test concepts, which suffice with logic, fail miserably when applied to memories.

It has been said that "memory testing is simple." The fact is that memory testing is logistically simple. Accessing one memory location in a sea of other memory locations is as simple as selecting a set of x-y coordinates. The complex part of memory testing is the numerous ways that a memory can fail. These numerous ways, also known as fault models, drive a myriad of patterns to test not only the cells but the peripheral circuitry around the memory cells as well. Understanding the correct fault models requires understanding the memory design, since different designs have different fault models. Once the appropriate fault models are recognized then the appropriate patterns and test strategies can be selected. This book will help the reader understand memory design, comprehend the needed fault modeling, and generate the appropriate test patterns and strategies. The design information contained herein provides a broad overview of the topology of memory circuits. The test information discusses pattern and associated test issues for the various memory types and fault models. The self-test information covers styles of self-test logic along with their recommended applications.

1. WHAT IS A MEMORY, TEST, BIST?

A memory is a means for storing computer information. Most often this storage is in the form of ones and zeros. These ones and zeros are stored and either retrieved or manipulated. The simplest block diagram for a memory is shown in Figure 1-1. Most memories can store data inputs to some location. The location is selected based on an address input. That same address is utilized to recall the data from the location. The information comes out of the memory in the form of ones and zeros after it has been evaluated by sense amplifiers. This is the simplest concept of a memory but one which is useful when understanding the detailed components of each of these structures, which differ from memory type to memory type. Certain memories do not have some of the blocks shown in Figure 1-1. Other memories have vastly enhanced blocks that bear little resemblance to the blocks named here.

The primary memories which are of concern in this text are embedded memories. Stand-alone memories have their own unique problems which are most often a function of the interface to the chip and not a function of the memory specifically. By examining embedded memories, the essence of the memory type is not obscured by a challenging set of off-chip timings and voltages.

A test for a memory involves patterns. The patterns are a sequence of ones and zeros chosen for a specific memory type. These patterns are applied at a set of environmental conditions, namely temperatures, voltages, and timings that aids in detection of defects. These environmental values are normally determined empirically for a given memory and processing technology.

A built-in self-test or BIST is the means for testing an embedded memory without the need for significant stimuli or evaluation from off chip. The BIST should be tailored to the memory being manufactured and is an enabler of high quality memories. It applies the correct pattern stimuli and does the correct evaluation for a given memory. These are quick definitions which will be fleshed out in detail throughout the remaining chapters.

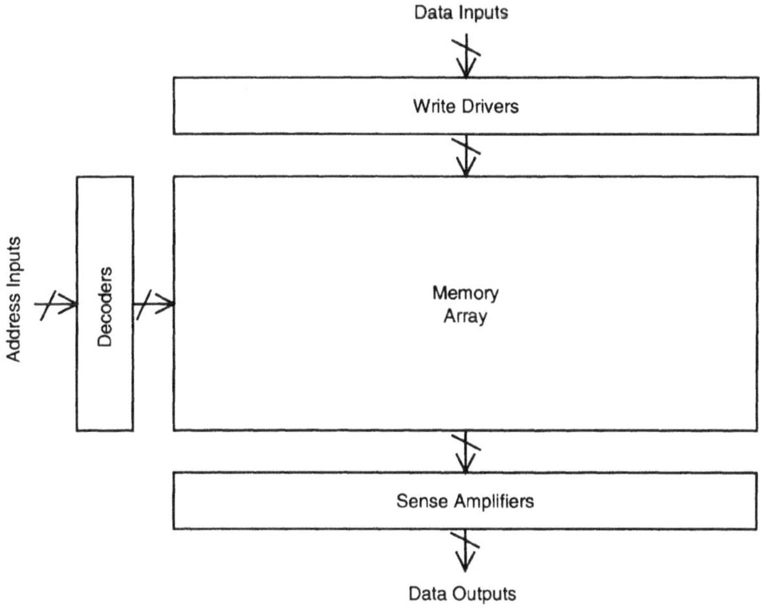

Figure 1-1. Simple block diagram for a memory.

2. THE UBIQUITOUS NATURE OF MEMORIES

No discussion of memories would be complete without a mention of Moore's Law [1,2,3,4]. In the mid 1960s, Dr. Gordon Moore stated that the number of transistors on a chip would double every year. In the mid 1970s, he revised the trend downward to a doubling every 18 months. This amazing trend has continued through time and is reflected in Figure 1-2. The number of transistors is directly proportional to the number of bits on a chip. In the case of a DRAM, the number is virtually identical since a single transistor corresponds to a memory cell, if peripheral circuitry is not considered.

In recent years, if a technology pundit wants to say something "provocative" or "radical" they start out by saying that Moore's law is dead. They then go on to state why the continued growth trend is impossible for their own favorite pet reasons. The growth trend, however, continues on unabated. Apparently, Dr. Moore now has another law. He states that the number of predictions that Moore's law is dead will double every 18 months. This second law may, if anything, be on the low side. Nonetheless, his first law is what concerns the readers of this book.

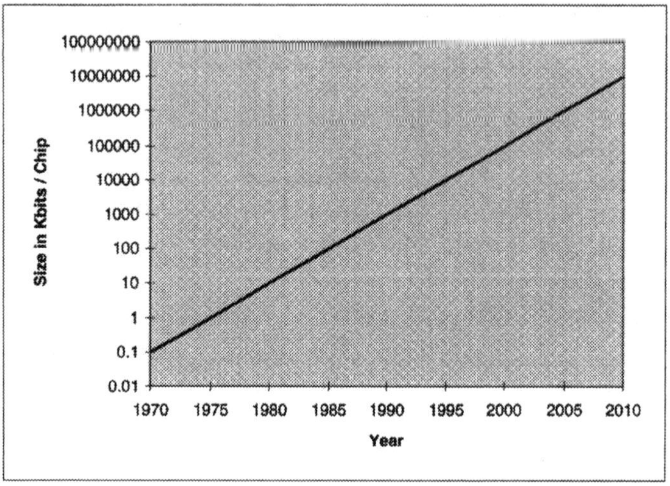

Figure 1-2. Diagram showing Moore's law.

The continuing trend for more and more memory on a chip should come as no surprise to the readers of this book. In fact, the appropriate response from a design and test perspective is simply a polite yawn. More bits of memory mean more decoding circuitry and more data inputs and outputs (I/O). It may mean a slight increment in test time to cover the additional bits. More bits of memory do not cause a shockwave of impact to the design and test communities. The greater memory density does create havoc in the processing community since it drives smaller feature sizes with their corresponding tighter tolerance for feature size variability. Further, it drives tighter limits on the number of tolerable defects in smaller and smaller sizes on the manufacturing line, since these smaller defects can still cause memory bit failures. That said, the greater density of memory bits does create stress within the industry but does not do so for the designers and test engineers.

3. THE COMPLEXITY OF MEMORIES

The presence of memories everywhere does, however, create serious challenges for the designers and test engineers. Logic is making its way into memories such as is the case of comparators in certain memory structures. Logic is placed in memories when it is the *logical* or appropriate design point. Logic circuitry that benefits from having direct access to a wide memory data bus only makes sense to be included as part of the memory proper. This logic is very regular in structure and also requires very high performance. Thus, it is custom designed similar to that of the memory circuitry itself. From a design point of view this typically involves wide-data high-speed circuit techniques. From a test point of view, it means that testing of this logic is required, which must be done through memory accesses. If this logic within a memory boundary is tested by means of logic test, the problem becomes a sequential test challenge, which is solvable but not simple. Logic test techniques are very effective at covering random logic but require special effort to test regular logic, such as that of logic embedded with memories [5]. The memory test techniques provide inadequate coverage of faults in the logic since the regular memory patterns are tailored for memories and not logic. There is some opportunistic fault coverage in the logic but it is small by any measure. The addition of logic to memories drives a significant complexity factor in design and especially test arenas.

Key point: The growth in memory complexity is more challenging than the growth in density.

Memories are becoming more deeply embedded. Memories formerly were on stand-alone memory chips. Testing of these memories was accomplished by memory testers applying through-the-pins test patterns. Later, a single memory was embedded with logic on a chip but all of the memory I/O were connected to chip I/O. Then a few memories were embedded on a single chip. Recently, the sheer number of memories on a single chip has become daunting to even consider. It is not unusual to hear of 40 different memory designs on a single chip with many more instances of each memory. Hundreds of memories can be contained on a single chip. Having direct chip I/O access for most of these memories is impossible. Thus, accessing these memories to apply appropriate test patterns requires considerable thought. By examining a typical microprocessor chip photograph it can be seen that much, if not most, of the chip real estate is covered with memories. In Figure 1-3 a photograph of an IBM PowerPC 750CX microprocessor wafer is shown. While this photograph is pretty, by examining a single chip location in detail, the level-two cache stands out. The smaller level-one caches can be seen as well, along with numerous other memories across the chip. Wherever a regular structure is seen, this feature can be attributed to a memory. Random logic appears as a rat's nest of wiring, connecting all of the various gates in the silicon. A memory, though, appears as a regular, attractive grouping of circuitry. Some have referred to the comparison of memory images with random logic images as "the beauty and the beast" with the memory obviously being the beauty.

Figure 1-3. Photograph of a PowerPC 750CX microprocessor wafer.

Key point: Memories often take up most of a chip's area.

While everyone realizes that memories are present in microprocessors and assume that the appropriate design and test challenges are being met, the fact is that memories are found in many other places as well. Personal digital assistants (PDAs), also known as pocket or handheld computers, have the amount of memory they contain as one of the first items in any advertisement. Cell phones contain significant memory, as do digital answering machines and digital recorders. As memories enter new locations, design and test challenges are created to meet the specific application needs as well as the processing technology.

The number of types of memory is becoming truly impressive as well. In the past there were dynamic random access memories (DRAM) and static random access memories (SRAM) as the workhorses. Now content addressable memory (CAM), both standard and Ternary (TCAM), is present in many chips. Further division of CAMs come as they can be composed of either static or dynamic style memory cells. Multi-port memories in numerous sizes and dimensions seem to be everywhere. There are two, four, six, nine, and higher dimension multi-port memories. There are pseudo multi-port memories where a memory is clocked multiple times each cycle to give a multi-port operation appearance from a single memory access port. Further, there are a number of memory entries that are non-volatile, such as Flash, EEPROM, FeRAM, MRAM, and OUM memories.

The number of process technologies in which memories are designed is quite large as well. There is the standard complementary metal oxide semiconductor (CMOS), which is the primary technology. A modification of this is the silicon-on-insulator (SOI) technology. Further, there is the high-speed analog Silicon Germanium (SiGe) technology, which now needs memories for numerous applications. Each quarter a new technology seems to emerge. These include "strained silicon" and "silicon on nothing."

Memory Types

- **Access/Storage method**
  - SRAM
  - DRAM
  - CAM
  - ROM
- **Form factor**
  - Embedded
  - Stand alone
  - With / without redundancy
- **Technology**
  - CMOS
  - SOI
  - FeRAM
  - MRAM
  - Ovonic
  - Flash
  - SiGe
  - GaAs
  - Copper

Figure 1-4. Summary of memory types, technologies, and form factors.

Each memory, when large enough, requires redundant elements to ensure significant yield can be produced. Testing the memories and allocating redundant elements further complicates the memory scenario. Each of the complexity factors listed above is driven by real and significant needs for performance, functionality, or manufacturability. Design and test engineers must face and meet these challenges. Figure 1-4 breaks out these challenges into categories of memory access, technology, and form factor. Examining these gives a quick glimpse into the complexity faced by the design and test engineers.

4. IT WAS THE BEST OF MEMORIES, IT WAS THE WORST OF MEMORIES...

Dickens' book, *A Tale of Two Cities*, describes two countries, two cities, and two socio-economic classes. For some it was the best of times but for many more it was the worst of times. In a far different way the same memory can be the best of memories or the worst of memories. There are multiple parameters which define the specifications a memory must face. Three handy continuums are performance, power, and density. Certain memory types are ideally suited for density, such as DRAMs. Certain memories are ideally suited for performance, such as SRAMs. Certain applications require lower power, like that of a personal digital assistant (PDA) while some require high performance, like that of a cache on a high speed microprocessor. There are other parameters that lend themselves not to continuums but rather to digital yes/no results. One example of this is retaining information even after a chip has been powered off, as is required for smart cards. Another need is being able to request data, given that one knows part of the data field, rather than an address, for which a CAM is well suited. Another "digital" specification is being able to fix soft errors, as error correction codes or ECC allows. This specification, however, is driven by the continuum of desired reliability.

From this discussion it can be seen that certain memory types are ideally suited for certain specific applications. Customers don't care whether a specific type of memory is used; they know what their application needs are. The role of memory designers and of memory test engineers is to understand the need, correctly select the memory type, and then test it thoroughly to ensure that the customer's application runs flawlessly, or at least as flawlessly as the system software allows. In other words the memories need to fully function within the application specification.

As a result the customer will never realize that "it was the best of memories" because they won't need to even think about it. However, if the application uses "the worst of memories" it can be assured that the customer will perceive it. No one wants to be working with or working on "the worst of memories." Certainly no customer wants to have the "worst of memories." Yet using a great memory in the wrong application will result in the "worst of memories." Customers hate to wait for a slow search, for their word processor to scroll, or for their PDA to scan the updated calendar.

Using the wrong memory will only exacerbate their sensitivity to this slowness. Figure 1-5 represents the continuum of power, density, and performance showing the relative locations of several memory types. There can even be variations in the chart placement for one specific type of memory. For example, an SRAM can be designed for low power applications by trading off performance. Likewise, one memory may have low power at standby conditions while another may have low power for active conditions. It is important to understand the memory and the specific application requirements in detail.

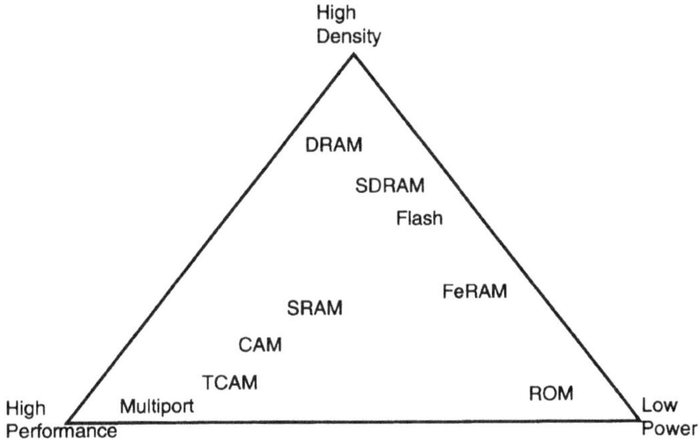

Figure 1-5. Diagram of three-dimensional continuum.

Key point: The memory type must be matched to the application.

5. TESTING: BITS IS NOT BITS

Some years ago one fast food company criticized its competitors with an advertisement campaign. Its competitor's product, it turned out, was made from "pulverized chicken parts." The ad went along the lines of a customer asking a kid behind a counter, "What kind of parts?" The response was a shrug of the shoulders and the statement that "parts is parts," meaning that the exact type of part did not really matter. All would agree that certain parts of the chicken are more appetizing than others. Many treat the issue of memory testing with not a "parts is parts" perspective but rather a "bits is bits" mentality. The testing of one type of memory, however, should be radically different from testing another type of memory.

On examining Figure 1-6, one should plainly see that a typical SRAM cell has six transistors with active pull-up and pull-down devices. A defect-free cell will retain its data as long as power is applied and nothing unforeseen happens to the memory. A typical DRAM cell, on the other hand, has one transistor and one capacitor. Since there is no active path for restoring charge to the capacitor, where the cell's data is stored, charge constantly leaks from the cell. The DRAM cell must be regularly refreshed. Thus, the operation of an SRAM and a DRAM is inherently different and the design is hugely different. This difference means that the testing of these two types of memory should be inherently different as well. To put it more simply, bits are not bits. Just as a tractor, a tank, and an automobile, though they all transport people require different testing to ensure quality, different types of memory bits require different testing. Even when considering just automobiles the testing would be quite different. An economy class car, a sports car, and a limousine would be expected to be tested in different ways prior to being received by the customer. So too would memory bits in different circuit topologies require different test patterns. The various types of memories, along with their bit cell designs, will be covered in later chapters so that the best test patterns and strategies can be applied to facilitate high quality testing, resulting in high quality memories.

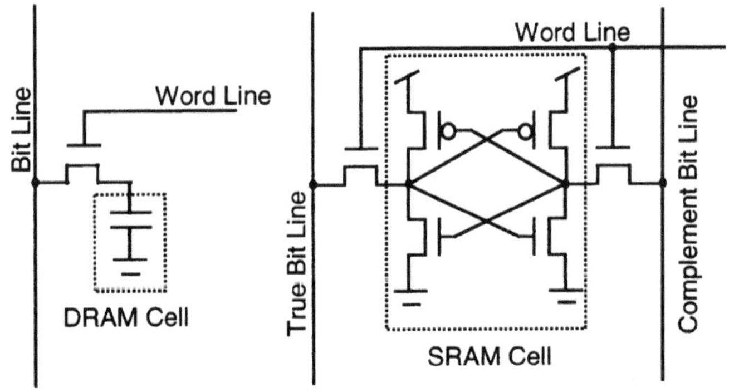

Figure 1-6. DRAM and SRAM Cell Structures.

Key point: Different types of memories require different types of tests.

6. BEST BIST OR BUST: THE JOURNEY TOWARD THE BEST SELF TEST

Proper self test of memories involves the understanding of memory design and the understanding of memory test. By combining these concepts together a comprehensive tester can be built into the chip itself. A built-in self-test or BIST engine directly applies the patterns to embedded memories, which typically would be applied with external test equipment to stand-alone memories using through-the-pins techniques. The BIST further does evaluation of the memory outputs to determine if they are correct. Beyond just applying patterns directly to the memory inputs, the BIST observes the memory outputs to determine if it is defect free.

The concept of a BIST is amazing when one thinks about it. External automated test equipment (ATE) are large expensive items. Instead of using these to provide stimulus and evaluation, they only provide clocking to the BIST and thus can be much simpler. The BIST engine handles the stimulus and observation functions while occupying but a small portion of chip real estate.

There are some very real advantages of using a BIST as opposed to ATE. First of all, it normally is the only practical way. Memories have become very deeply embedded within chips. Deeply embedded means that there are more memories on a chip and these are buried deeply inside the functional logic of the chip. It is no longer the case that a single memory is on a chip and that the memory I/O can be brought to the chip's I/O connection points. Now there can be 100 or more individual memories on a chip. These memories are at different logical distances from the chip I/O. Even if an attachment from a memory could be made to chip I/O, the memory's I/O would be loaded down with extra capacitance, thus encumbering the functional performance. Getting patterns to an embedded memory from an ATE cannot be accomplished at the same rate of speed as the memory can function. On-chip memories can function at 2 GHz or higher speeds. Testing, by applying patterns cycle after cycle at speed, cannot be accomplished by ATE. A BIST does not care how deeply embedded the memory is. Further, a BIST can be designed that will run at any speed a memory can run at. These reasons cause the quality of test, and thus chip quality, to be better through using a BIST.

There are some significant conceptual and aesthetic reasons that a memory BIST is superior to ATE test of embedded memories. Often times testing can be relegated to the last thing considered. A design can be completed and then "thrown over the wall" to the test engineers. Their responsibility becomes one of figuring out how to test an ungainly design.

Most organizations have come to realize that this approach is a recipe for disaster and have learned to consider test early in the design phase of a project. Nonetheless, due to schedule and headcount pressures there is always the temptation to revert to a "We'll design it and you figure out how to test it" mentality. Obviously the customer's function is paramount but quality is also paramount and is driven by the test capability and test quality for a given chip. Having a memory BIST pushes the test into the design phase, since the BIST engine must be designed into the chip. Thus, the test and design teams are forced to work together; BIST forces this to happen.

Key point: Test must be developed concurrently with the design.

There is the issue of yesterday's transistors versus today's. ATE must employ older technology since it takes time to design and produce an ATE system. BIST, however, employs the latest transistors to test the latest memory circuits. Because of this fact a BIST design need never limit the speed at which memory design is tested. BIST can apply at-speed patterns cycle after cycle until the complete memory is tested resulting in a short test time. Sometimes, test of embedded memories with ATE is accomplished through scan access. In this situation a pattern is generated on the external tester and scanned serially into the chip until the correct data, address, and control inputs are lined up with the memory latch inputs utilizing the scan chain. The memory is then clocked. Any read data is then scanned out into the tester for comparison with expected values. For each memory test cycle an extensive scan procedure is required. This causes the test time to be extraordinarily high since scanning the correct values into place for a single cycle can take four orders of magnitude longer than clocking the memory operation for that single cycle.

Further, the at-speed test of a BIST provides a better quality test. Fast cycling, operation after operation will generate more noise on chip and will allow more subtle defects to be detected. Any defect in the pre-charge or write-back circuitry might be missed with slower ATE scan testing but with at-speed BIST testing they can be caught.

A slower test drives more test time and more test time means higher test cost. When some see that testing can run up to 50% of the chip manufacturing cost [6,7], they logically want to reduce cost wherever possible and test time is an obvious point. Designing a memory BIST does mean some development expense for the project. However more test time, i.e. not having BIST, means greater cost for every individual chip produced for a lower quality test.

There is much discussion in the test literature on test re-use. Often a chip needs to be tested before wafer dice, after package build, and again after package burn in. This package is then incorporated onto a card, which needs to be tested. The card goes into a system, which needs to be tested. Then the system goes to the field, for customer utilization, and it needs to undergo periodic test, normally at each power on, to ensure continued proper operation. All of these tests can be generated individually but the inefficiency is great. Utilizing a BIST at all these levels of assembly means that a high quality test can be applied at each step in the manufacturing and use process. Further, it means that the effort spent to determine the best test during the design phase of the project is re-used over and over again, thus preventing wasteful development of test strategies at each level of assembly.

Lastly, it is inherently elegant to have a chip test itself. Having a complex circuit, such as a memory, test itself and thereby solve the test challenge while solving the customer's test problem is quite attractive. BIST is a clean solution.

Key point: For embedded memories, BIST is the only practical solution.

In conclusion, there are numerous reasons that make BIST testing of a memory very attractive. Certainly, for virtually all embedded memory applications, BIST is the only practical and logical solution. Beyond this, however, the right BIST must be utilized for testing of the memory. The BIST must be tailored to apply the patterns that best identify defects and thereby make the highest quality end product. A BIST that has a weak pattern set may be in the right physical location and may be fast, but will result in shipping defects and thus dissatisfy customers. The BIST must be the right one for the memory and the best one that can be developed. That results in the best quality, the best test, and the shortest test time. In short it provides the best BIST.

7. IGNORANCE IS NOT BLISS

Memories, though they simply store ones and zeros, plainly have significant complexity to them. The case has been made that the various types of memories have significant differences in them and require differences in test. There are two points where knowledge is the key to having satisfied customers. It is a given that they will not be satisfied if their memories are failing, thus the designers and test engineers must ensure that this event does not happen.

The first point where knowledge is key is knowing when a memory is defective. That means that defective memories must be identified and not allowed to pass final manufacturing test. A simple test is indeed still a test.

Naivety would allow a test to be performed and the results to be taken for granted. If a simple, one might even say *dumb*, test is executed a memory can pass this test. That certainly doesn't mean that the memory is good. It is frightening to hear someone say that the memory "passed" and then obviously not understand what sort of test was entailed or how thorough it was. Thus it is obvious that the statement "ignorance is bliss" is not the case when it comes to defective memories. Not knowing that a memory is defective allows it to be shipped to the customer. That customer will not allow the supplier's life to be blissful once their parts start failing in the field.

There are times too numerous to count where a memory nightmare occurs. This nightmare starts with a large amount of memory fallout. The memory test results are pored over to determine the exact type of failure. During this examination it becomes obvious that the memory tests are inadequate. It is then the memory expert's responsibility to tell the parties involved that, not only is the memory yield low but the test was insufficient and the real good-memory yield is even lower than it was already purported to be. Ignorance, i.e. testing without knowledge, may result in an artificially high yield but the result will be disastrous. Simply put: a memory that passes test is only as good as the test applied! Having an adequate memory test is key to identifying and culling out the defective memories.

Key point: A defective memory can pass a poor quality test.

In order to identify defective memories, high quality test strategies must be employed. Obtaining these high quality strategies requires a good understanding of the memory design and a good understanding of memory testing in general. This is the second point where knowledge is key to having satisfied customers. Ignorance in either of these areas will lead to a poor quality memory. Thus it is possible for someone to be ignorant with respect to memory design or ignorant of memory testing and then, in turn, be ignorant of the fact that defective memories are being shipped with "good" labels on them. Ignorance is definitely not bliss.

8. CONCLUSIONS

Design and test are considered jointly in this book since knowledge of one without the other is insufficient for the task of having high quality memories. Knowledge of memory design is required to understand test. An understanding of test is required to have effective built-in self-test implementations. A poor job can be done on any of these pieces resulting in a memory that passes test but which is not actually good. The relentless press of Moore's law drives more and more bits onto a single chip. The large number of bits means that methods that were "gotten away with" in the past will no longer be sufficient. Because the number of bits is so large, fine nuances of fails that were rarely seen previously now will happen regularly on most chips. These subtle fails must be caught or else quality will suffer severely.

Are memory applications more critical than they have been in the past? Yes, but even more critical is the number of designs and the sheer number of bits on each design. It is assured that catastrophes, which were avoided in the past because memories were small, will easily occur if the design and test engineers do not do their jobs very carefully.

In the next few chapters an overview of the various memory designs will be provided. The section after that will provide a summary of memory testing. The last section will detail the key factors in implementing good self-test practices.

Chapter 2

Static Random Access Memories

Design & Test Considerations

"Memories. You're talking about memories ..." – Harrison Ford in the movie *Blade Runner*

Static random access memories (SRAMs) have been, are, and will continue to be the workhorse of memories. While there are more DRAM bits worldwide, especially when considering embedded memories, there are a larger total number of SRAM memories. SRAMs are subtly inserted into countless applications. SRAMs were the first memories produced. SRAMs are fast and are utilized where the highest speed memories are required, such as the L1 caches of microprocessors. They can be designed for low power application requirements. Further, they retain their data until the power is removed or until the data state is modified through writing to a cell location. Of all semiconductor memories, the SRAM is the easiest to use. There is no required refresh of the data, accessing is performed by simply providing an address, and there is only one operation per cycle, at least for a one-port SRAM.

This chapter will provide the design background for the remainder of the book. The SRAM will be used as the model through which all other memories are examined. The memory cells, pre-charge circuits, write drivers, sense amplifiers, address decoders, and redundant elements will all be examined. When other memories are discussed in this book the differences to SRAMs will be noted. Many times the circuits will be the same as that for SRAMs, in which case the reader can simply refer back to this chapter to better understand the design. The design schematics provided are examples; many subtle variations are possible. Once a proficient understanding is obtained of the basic design, test strategies will become more obvious. As design variations are encountered, nuancing the test patterns will follow logically.

1. SRAM TRENDS

The base of any memory is a single cell into which data is stored. A static random access memory is no different. The cell must be small. Since the cell is replicated numerous times, it is highly optimized in each dimension to be able to pack as many cells together as possible. Further, due to process scaling, the area of cells rapidly decreases over time [8]. The trend for SRAM cell size, as a function of time, is shown in Figure 2-1.

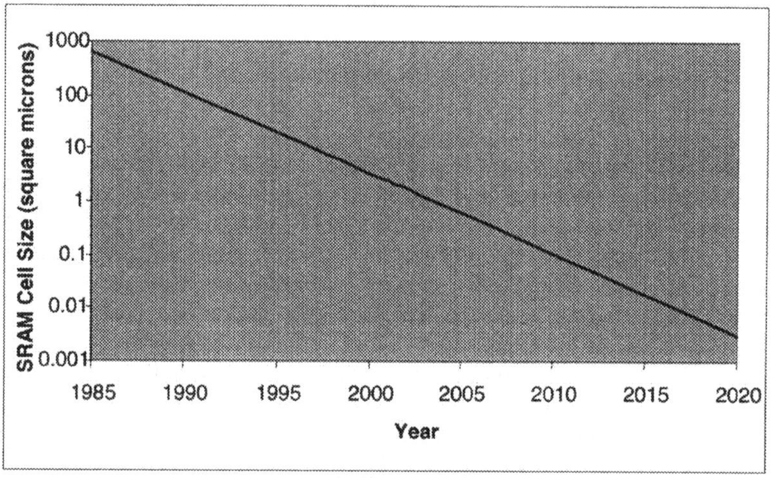

Figure 2-1. SRAM cell size trend.

Another representation of the reduction in SRAM cell size is shown in Figure 2-2. The numbers for this chart come from the SIA roadmap [9]. The values on the y-axis show the six-transistor SRAM cell area factor trend with overhead.

With this constant downward trend in cell size and with the insatiable customer demand for more performance and memory, the number of bits put on a chip continues to increase. Figure 2-3 shows an analysis of the SRAM density increase over time. These are SIA roadmap based numbers showing the increased number of SRAM transistors in a square centimeter over time.

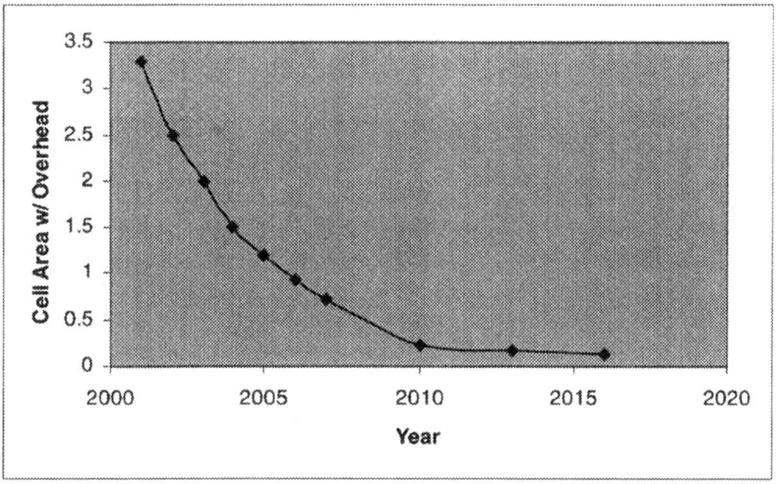

Figure 2-2. SIA roadmap of cell area with overhead.

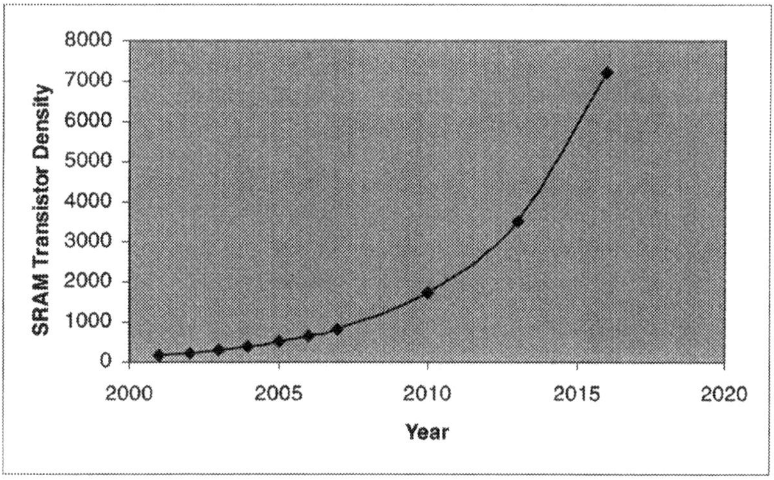

Figure 2-3. SRAM density in millions of transistors per square cm.

Since performance is the paramount factor in SRAMs, the memory access time continues to decrease. Furthermore, memories are having a greater number of data inputs and data outputs. This means that the overall performance, measured in terms of bits of access per second is rising dramatically. Figure 2-4 shows this increase in SRAM performance

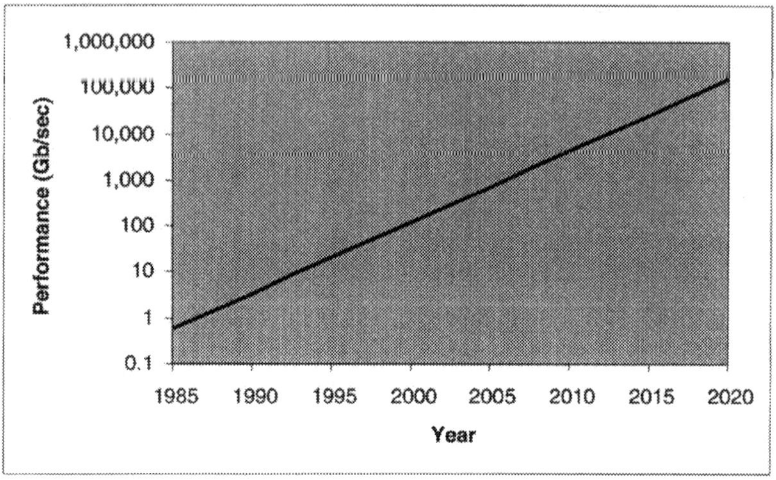

Figure 2-4. SRAM overall performance trend.

2. THE CELL

Each SRAM cell must be easy to write and yet be stable, both when in a quiescent state and when it is being read. The cell must store its binary data regardless of the state or the operations being performed on its neighbors. The standard SRAM cell is made up of six transistors as shown in Figure 2.5. There are two pull-down devices, T2 and T4, two transfer devices, T5 and T6, and two pull-up devices, T1 and T3. In Figure 2.6 an alternative is shown with only four transistors, where the two pull-up transistors have been replaced with resistors. The four-transistor configuration was occasionally used in stand-alone memories. Today most SRAM cells are of the six-transistor variety, especially the embedded ones. These have lower quiescent power and greater soft error resistance.

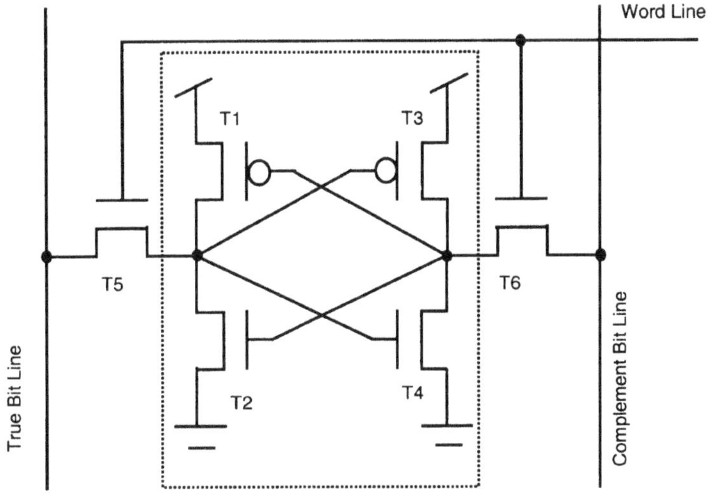

Figure 2-5. Six transistor SRAM cell.

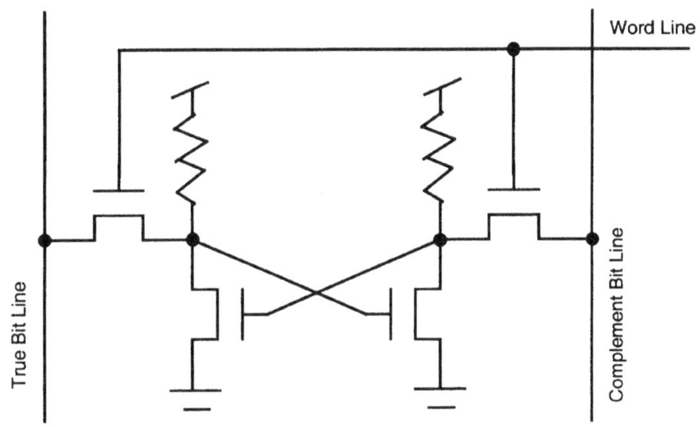

Figure 2-6. Four transistor SRAM cell.

There are numerous ways that an SRAM cell can be laid out [10]. Figure 2-7 shows one simple layout for a six-device SRAM cell [11]. The two transfer NFETs are at the bottom with the horizontal polysilicon (or poly for short) shape making up the word line that is lightly shaded. Two black bit line contacts are below where the cell is attached to the true and complement bit lines. Above the word line, in the center, is the ground contact. At the top of the figure is the Vdd contact. The lighter and darker shaded shapes in the center make up the cross-coupled latch. The large unfilled shapes are the diffusions. Wherever the poly crosses the diffusion a transistor is formed. The two pull-up PFETs are at the top and sit in an Nwell, which is not shown. For this illustration it is assumed that the chip is formed on a P-minus epitaxial layer and that no Pwell is required. Alternatively, the four NFETs at the bottom of the figure may be incorporated in a Pwell for a twin tub process.

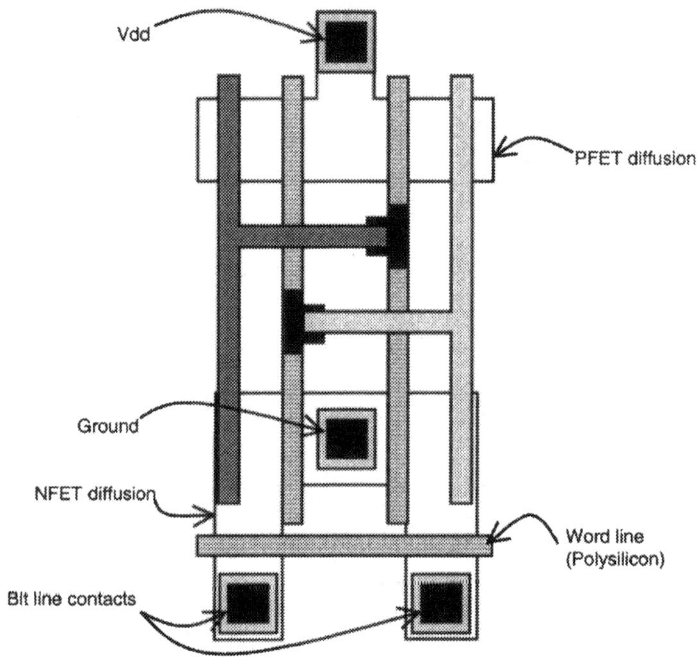

Figure 2-7. One layout for a six transistor SRAM cell.

An alternative layout incorporates two ground contacts and two Vdd contacts. An example of this layout structure is shown in Figure 2-8. The area of a memory cell is the primary factor in the overall area of an embedded memory or a memory chip. Thus the layout of a single cell is very carefully optimized to shave off every possible piece in both the x and y dimensions. Normally the layout of an SRAM cell along with the associated analysis will take several months of effort. While the layout in Figure 2-8 may appear to be larger than the previous six-device layout it should be noted that the Vdd and ground contacts can be shared with adjacent cells. It is possible, and even typical, for a single Vdd contact to be shared between four adjacent cells, thus saving significant area. Also, since there are no contacts in between the cross-coupled FETs, the spacing between these shapes can be driven to very small values. Often, bit line contacts are shared by a pair of cells vertically adjacent to one another along a column. Again, if a single bit line contact becomes excessively resistive [12] then not one cell but two will fail. Since reading a cell involves the cell pulling down either the true bit line or the complement bit line low, a resistive bit line contact causes one of the two data types to fail on these two cells. Thus, these two vertically paired cells with a defective bit line contact may be able to store and read either a "1" or a "0" but not both. Furthermore, a resistive bit-line contact degrades the writing of the cells more than it degrades the reading of the cells.

Because the SRAM cells in figures 2-7 and 2-8 are laid out differently, they fail differently as well. One cell layout style is sensitive to different manufacturing defects and fails differently from other cell layout styles. This means that different fault models and testing patterns should be used for these different designs.

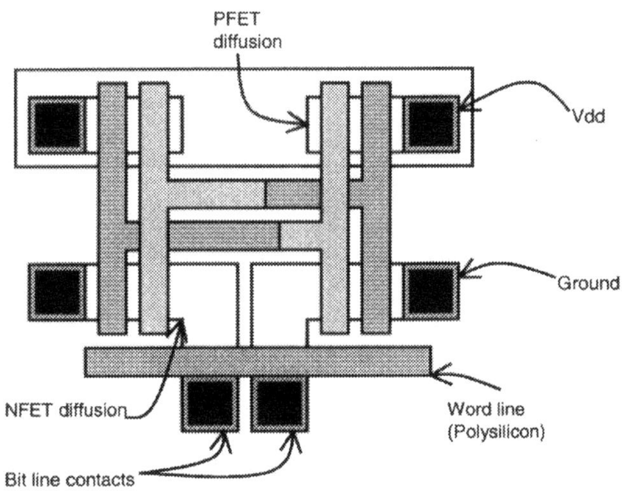

Figure 2-8. Alternative layout for a six transistor SRAM cell.

For example, the cell in Figure 2-8 has separate ground contacts. If one of these ground contacts is resistive then the cell can easily disturb since there is an imbalance between the true and complement pull-down paths. The cell with a single ground contact, if it is resistive, has common mode resistance to the true and complement nodes of the cell, thus causing the cell to retain most of its stability, even with the defective resistance. A second example defect is an open Vdd contact. For the cell layout in Figure 2-8, where a Vdd contact is shared by four adjacent cells, an open Vdd connection causes not one cell to fail but rather a group of four cells. Thus, even though the schematic for the two layouts is identical, the failure modes for the two layouts are different.

For any layout configuration, the cell stability is defined by the ratio of the strength of the pull-down transistor divided by the strength of the transfer device. This is known as the "beta ratio." Normally, the beta ratio is simply the width of the pull down device divided by the width of the transfer device. Equation 2-1 provides the calculation when the lengths differ between the pull-down and transfer devices. It should be remembered that the dimensions of concern are the effective widths and lengths, not the drawn dimensions. A beta ratio of 1.5 to 2.0 is typical in the industry. A beta ratio below 1.0 indicates that each time the cell is read, it is disturbed as well. For SRAMs, a defect free cell must have a non-destructive read operation.

$$\beta = \left(\frac{W_{eff}\, PullDownFET \,/\, L_{eff}\, PullDownFET}{W_{eff}\, TransferFET \,/\, L_{eff}\, TransferFET} \right)$$

Equation 2-1. Determining the beta ratio, which, defines cell stability.

An SRAM cell's stability can also be described by the "butterfly curve." Figure 2-9 shows an example curve with one node voltage being displayed on the x-axis and the complement node voltage being displayed on the y-axis. A larger box, which is contained inside the butterfly "wing" indicates a more stable the cell. The butterfly curve illustrates the stability of the four latch transistors inside the cell. To generate a butterfly curve, one node is forced to a potential and the complement node value is determined. While each cell node is forced, the complement node is read. This defines the two curves, which make up the butterfly curve. The curve flattens during a read operation [13], reducing the box size and so illustrating a reduced stability. A smaller box size during a read also correlates to a smaller beta ratio.

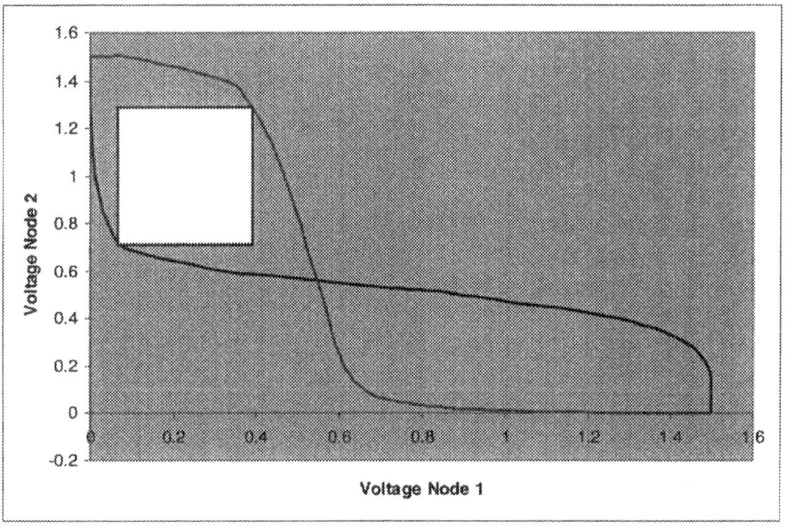

Figure 2-9. Example butterfly curve for an SRAM cell.

3. READ DATA PATH

The read data path can be examined by working outward from the cell to the memory data output. The cells of a memory are joined along a column. The apparatus for joining these cells together is a bit-line pair as shown in Figure 2-10. High speed SRAMs always use a pair of bit lines whereas small, lower performance SRAMs can have a single bit line for reading and another for writing. Further, it is possible to have only a single bit line for both reading and writing. With a single read bit line, a full or almost-full swing data value is allowed to develop on it during a read operation. The cells attached to these full-swing bit lines are typically larger in size to drive full logic values in reasonable time frames. The number of cells along a differential bit-line column is normally large, often being 256, 512, or 1024 in value.

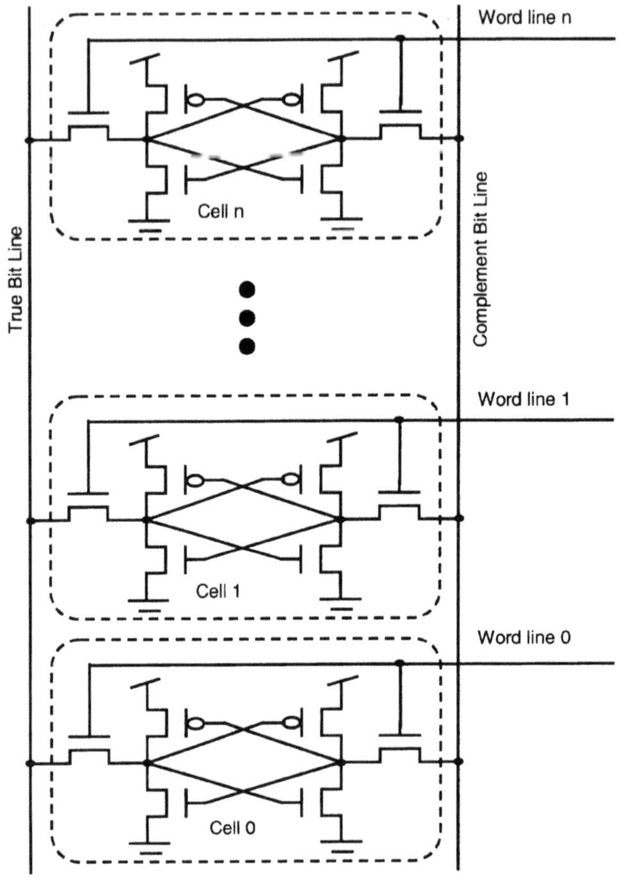

Figure 2-10. Bit-line pair attached to an SRAM cells along a column.

On typical high-performance SRAMs the bit lines only swing a small amount on a read operation. Such a read operation is initiated by an address being input to the memory and a clock going active. The word line corresponding to the input address goes high, in turn selecting a given row of cells. The word line turns on the transfer devices for a single cell in a column. Either the bit-line true or the bit-line complement is discharged through the appropriate transfer device, T5 or T6 respectively. If a "0" is stored in the cell being read, the true bit line is discharged and the complement bit line remains high. In the case of a "1" being read from a cell, the complement bit line is discharged and the true bit line remains high.

Only one of the two bit lines moves in value during any single read operation, and the bit line that changes in value does so by only a small amount. Due to the small excursion on one of the differential bit-line pair, there are some very real analog effects in memory operation.

Logical operation is typically represented as a "1" or a "0", however, the "1" or "0" stored in a cell is distinguished based on a small signal swing. Typically it may only be a 100 mV difference on a bit-line pair. Both bit lines are pre-charged into a high state. Most often this pre-charge is to a Vdd value but some designs pre-charge to a threshold voltage below Vdd. Pre-charging the bit lines to a threshold below Vdd is error prone and any differential between the bit line potentials seriously hinders the sensing capability since such small differences determine the correct "1" versus "0" value of a cell. It is therefore not a recommended practice [14]. Pre-charging to Vdd is easier and the norm in SRAM designs today.

Key Point: Analog effects in memories drive critical design and test issues.

The pre-charge to Vdd is most frequently accomplished by a three-transistor circuit, as shown in Figure 2-11. This circuit is sometimes referred to as a "crow bar". It forces each bit line to Vdd and also equalizes their potentials. There is a PFET pulling each bit line to Vdd and a third PFET connecting the two bit lines together. An alternative to this circuit leaves out the third PFET and simply has the two PFETs to pre-charge the bit lines to Vdd. During a read the pre-charge circuit is normally turned off for a column that is being read. For the columns that are not being read, the pre-charge is often left on. With the pre-charge in the on state and the word line being high on the unselected columns, the cell fights against the pre-charge circuit. The small amount of current consumed by this contention is usually very tolerable on a cell basis. The total power question is really one of consuming the flush through current between the cell and the pre-charge circuit or consuming the Cdv/dt current recharging all of the bit lines after a read. The cells that are fighting against the pre-charge circuit are said to be in a "half-select" state. Defect free SRAM cells have no problem retaining data in a half-select state since NFET transfer devices have such poor pull-up characteristics. The half-select state can actually be utilized as a feature to help weed out defective or weak cells. It should be noted that the bit line pre-charge signal is active low.

The bit lines are connected to a sense amplifier through isolation circuitry. An example isolation circuit is composed of two PFETs, as shown in Figure 2-12. The bit lines are isolated from the sense amplifier once sufficient signal is developed to accurately sense by the bit line isolation (ISO) signal going high. The reason the bit lines are isolated from the sense amplifier is to speed up the sense amplifier circuit operation. Bit lines are long with many cells attached to them. All of these cells load down the bit lines but the long metallization of the bit line forms even more load due to its large capacitance. The isolation circuitry shown in Figure 2-12 assumes that a single sense amplifier exists for each bit-line pair. For the case where multiple columns feed a single data out, as is normally the case for larger memories, the isolation circuit is replicated to form a multiplexer. This arrangement is referred to as a bit switch circuit. The appropriate pair of bit lines is attached to the sense amplifier, which correspond to the column address applied to the memory. Typical column decodes are two, four, or eight to one. They can also be 16 or 32 to one but this may involve another multiplexing stage after the sense amplifier. The exact decode width defines the column decode arrangement and therefore the aspect ratio of the memory. A four to one bit switch isolation circuit is shown in Figure 2-13.

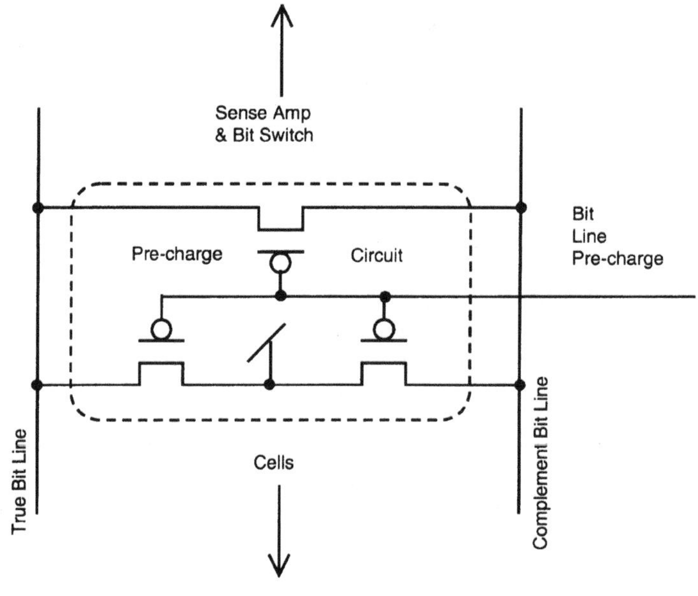

Figure 2-11. Pre-charge circuitry.

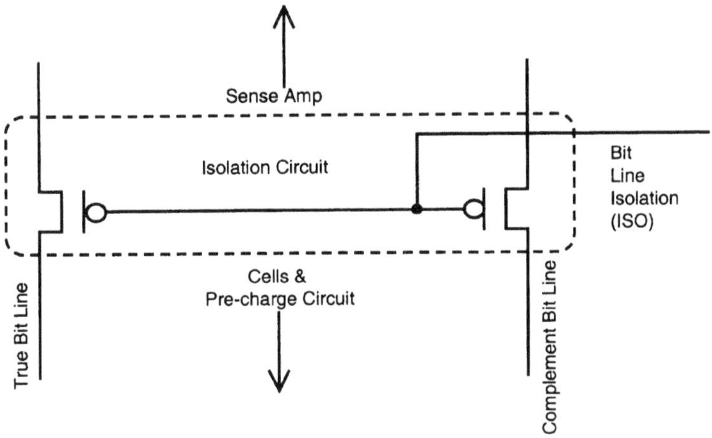

Figure 2-12. Isolation circuitry.

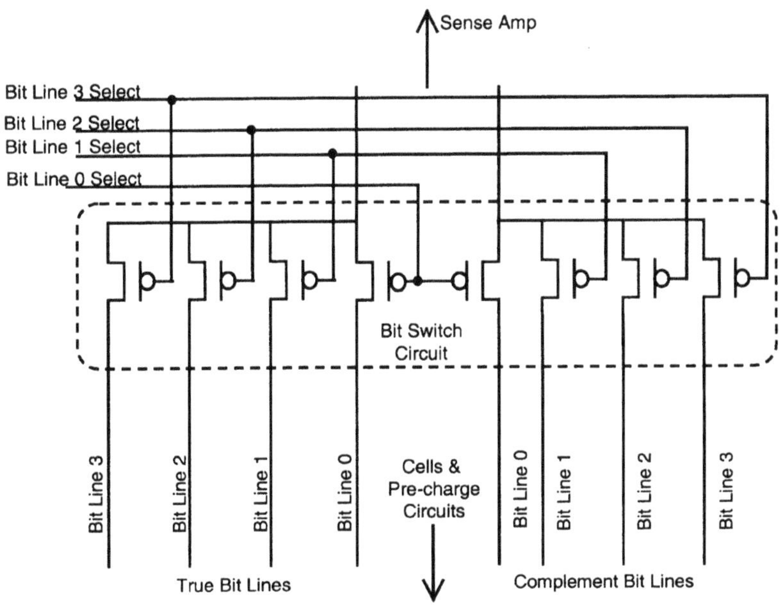

Figure 2-13. Bit switch isolation circuit.

There are many different types of sense circuitry. A latch type sense amplifier is shown in Figure 2-14. For this circuit the bit-line differential is applied to the drains of the four transistors forming the sense amplifier's latch. An alternative is to have a latch type sense amplifier where the differential signal is applied to the gates of the NFETs from the sense amplifier latch as shown in Figure 2-15. When this configuration is used a different bit line isolation circuit may be employed. Another alternative to the latch type sense amplifier is to remove the PFETs from Figure 2-14 [15]. When this circuit arrangement is used, the isolation circuit keeps the bit lines attached to the sense circuit and the bit lines hold the high node in an elevated state while the low node is actively pulled down by the sense amplifier. A second stage of sensing circuit is then normally employed to further amplify and latch the sensed result [16]. Often times a second stage of sensing is utilized to improve overall performance and latch the sensed result, regardless of the first stage's sense amplifier design style.

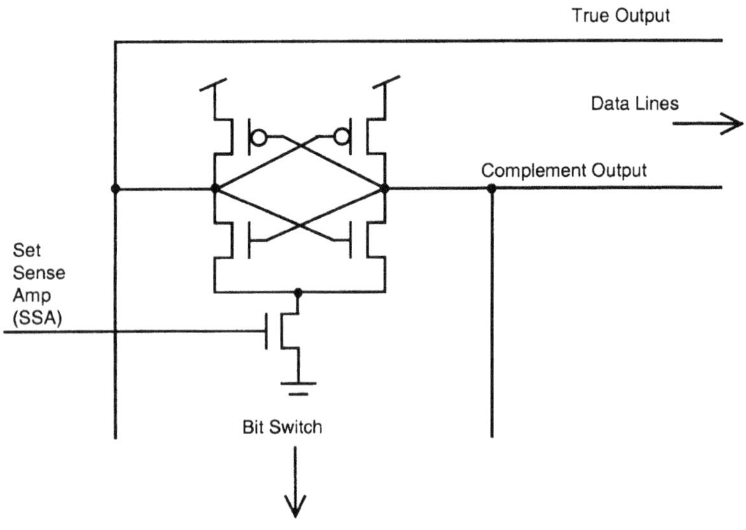

Figure 2-14. Latch type sense amplifier.

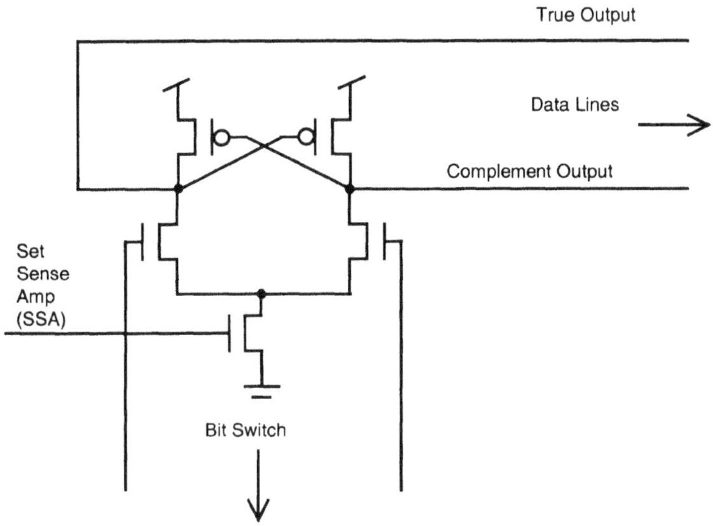

Figure 2-15. Latch type sense amplifier with differential being applied to the NFET gates.

For any of these latch type sense amplifiers, the sense amplifier is activated by the set sense amp (SSA) signal going high [17,18]. The differential in the sense amplifier is amplified to a full rail signal once the SSA signal is high. When there is only a single bit-line pair per sense amplifier, since the ISO and SSA signals are of similar phase, they can actually be a single signal. Once sufficient signal is developed into the sense amplifier, the bit line can be isolated and the SSA line causes the signal to be amplified and latched. Thus, SSA can drive the ISO input to the isolation circuitry when a single bit-line pair exists per sense amplifier. When multiple bit-line pairs feed a sense amplifier through a bit switch, the normal practice is to have the SSA signal go high slightly before the ISO signal goes high. It should be noted that when the ISO signal is brought up, both bit lines are coupled up via Miller capacitance. If the sense amplifier has started to set then the small signal developed on the bit lines tends not to be disturbed by coupling from the ISO signal. Further, the delay of getting data out of the memory is reduced by bringing the SSA signal in a little sooner. Even bringing SSA high will cause some coupling of the true and complement output nodes of the sense amplifier high, although this is more of a second order effect. The exact amount of sense amplifier signal developed can be considered the value at the point in time when the two nodes start to couple down as the sense amplifier starts to pull the nodes apart. Figure 2-16 shows an example set of waveforms. The cell's true node pops up indicating that the word line has gone high. The signal starts to develop on the true bit line. The complement bit line remains high. Signal stops developing on the sense amplifier true node when the ISO signal goes high. The SSA signal goes active causing the true sense amplifier output to go low. In Figure 2-16 the ISO and SSA signals have been purposely separated in time to illustrate their respective coupling effects on the bit-line pair. In actuality the ISO and SSA would be almost immediately adjacent in time.

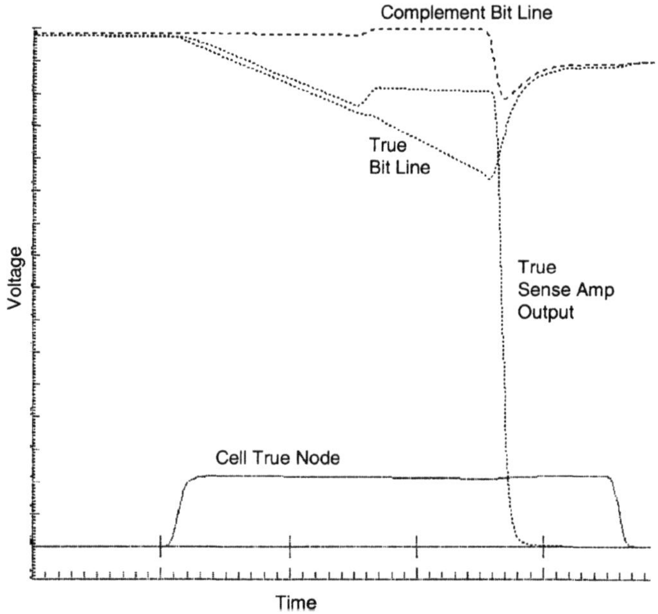

Figure 2-16. Example sense amplifier signal being developed and latched.

A current type sense amplifier may also be employed, an example of which is shown in Figure 2-17 [19,20]. This type of sense amplifier examines the bit line current flow to establish whether a "1" or a "0" is stored in the cell being read. A current sense amplifier does not operate like a latch type sense amplifier. A latch type sense amplifier locks in the data being read once the set sense amplifier (SSA) signal goes active. If more signal is generated on the bit lines or if the original differential was erroneous and the correct differential is established, the latched sense amplifier data does not change. It remains where it was at the time of SSA going high.

A current sense amplifier will evaluate when the SSA signal becomes active. The operation, however, is such that if more signal or the opposite differential signal starts to develop the current sense amplifier can correct itself [21]. Due to this difference, a current sense amplifier can evaluate more slowly for reading certain defective cells while most bits will evaluate more quickly. A latch type sense amplifier always evaluates at the same point in a read cycle and similarly provides its data output at the same time in every cycle.

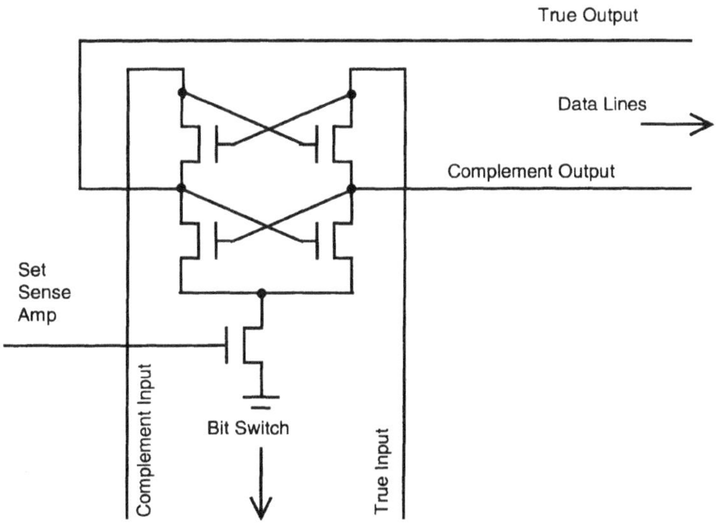

Figure 2-17. Current sense amplifier.

A key signal in the sense amplifier signal development is the set sense amplifier node. The timing of it is critical since it is used to determine when to turn on the sense amplifier and in many cases when to stop sensing the signal on the bit lines. Thus the timing is very carefully designed to track with the bit line signal development. If the SSA signal is poorly designed, then time is wasted as more than the needed signal develops into the sense amplifier. Worse still, the SSA delay can be short causing insufficient signal to reach the sense amplifier. This causes inconsistent results in the data being sensed [22]. The correct binary value can be in the cell but it can be sensed sometimes correctly and sometimes incorrectly. Thus, it is key that the SSA delay be very accurately modeled and designed robustly to track the bit line signal development.

Since the delay to SSA going active is so critical several design methods allowing good tracking have been developed. One method utilizes a dummy word line where the load on the dummy word line is similar to that of the normal word line, along which the cells are being read. Figure 2-18 shows a dummy word line with the cells tied off so as not to allow connection to the bit lines that pass over them. Other names for the dummy word line are standard word line, model word line, or reference word line. These names correctly indicate that this extra word line that is utilized only for timing.

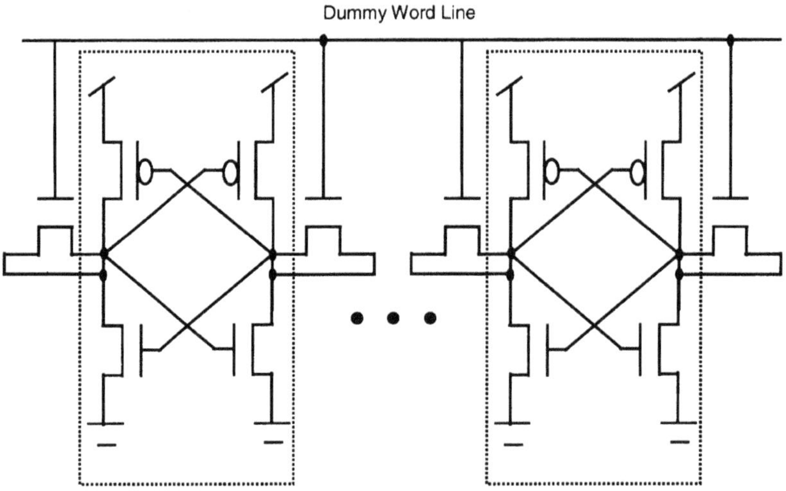

Figure 2-18. Dummy word line and its cell attachments.

An alternative to a dummy word line is a dummy bit line. Each time that a word line is brought high to read a cell, an additional cell is accessed which pulls down the dummy bit line. The end of the bit line is connected to logic circuitry that picks up this result and subsequently brings the SSA signal high. A third method involves simply including a series of inverters to provide the needed delay. Regardless of the method for delaying the setting of the sense amplifier the proper amount, it is critical that the delay tracks with signal development. As the process, temperature, and voltage applied to a design vary, the rate at which the signal on the bit lines develops varies. If the SSA delay does not track accurately with the signal development rate then the amount of signal on which the sense amplifier sets will be different for different conditions. If this variation exists then, at certain conditions, the correct data may be sensed while at other conditions the amount of signal will make sensing inconsistent.

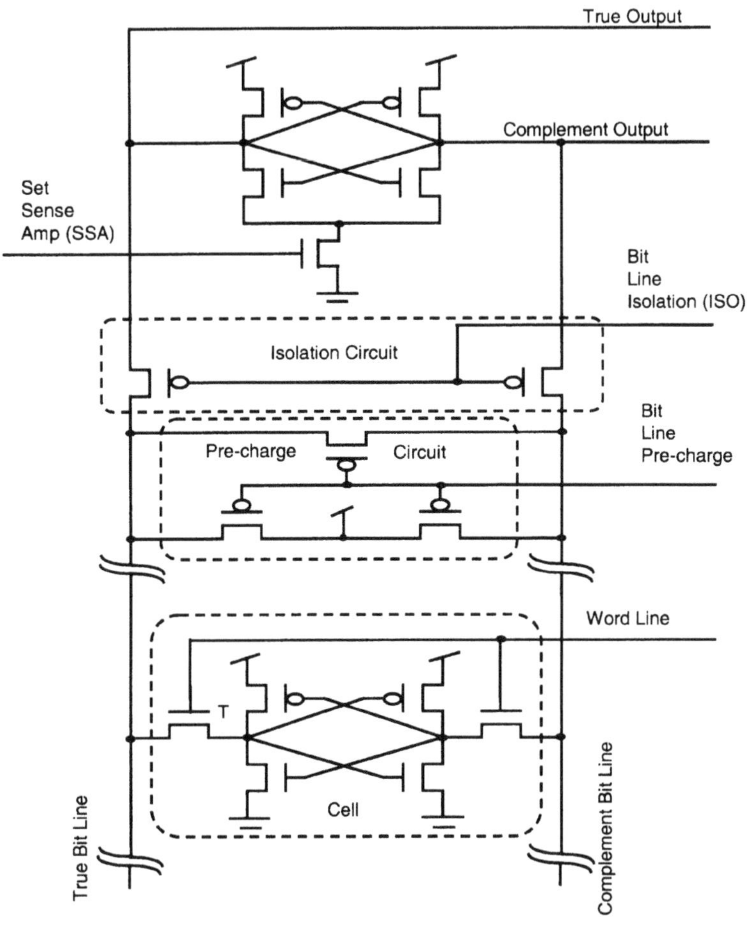

Figure 2-19. Read data path cross section.

Once the data output is latched in a sense amplifier, either in the first or second stage, it is provided to the output of the memory. The sense amplifier output can be multiplexed with other sense amplifier outputs, if there is the need for further column decoding. It can also be loaded into a latch, which will retain data into the next system cycle and beyond. The memory output can be driven across a tri-state bus for use by a processor or some other on chip component. Further, it may be driven to another chip by means of an off-chip driver. The number of possibilities is enormous and is not pertinent to the memory-proper design and testing. They are more a function of I/O design and therefore will not be covered in this text.

Figure 2-19 is a composite diagram of the cell, pre-charge circuit, isolation circuitry, and sense amplifier. These form a typical read cross section for an SRAM and are in many ways the key to understanding all of static memory testing.

Figure 2-20 shows the relative timings of the key SRAM signals. Depending on circuit topology these can vary but the signals are displayed here for comprehension purposes. The sense amplifier output signals show both the true and complement values. Note that one remains high while the other transitions low.

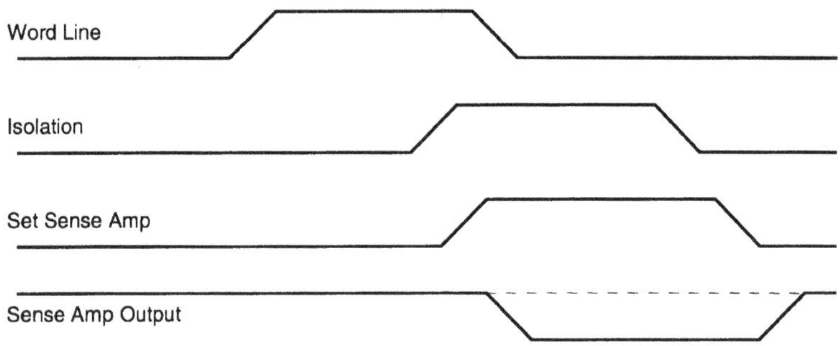

Figure 2-20. Typical read timing logic waveforms.

The various memory cells, isolation devices, bit switches, and sense amplifiers can have numerous subtle design variations. These should be understood prior to completion of the remainder of the design and certainly prior to completing the test strategy. Each transistor design difference can bring in new fault models, which need to be tested. Certain design differences also preclude other fault models and therefore eliminate the need for specific test strategies and patterns. The design styles presented here are intended to provide an appreciation of the memory design and facilitate the test of it. There is not a single golden test strategy that works on all SRAMs, since there are nuances in different designs that drive corresponding test differences. This reminder is placed here since the read data path has most of the subtle SRAM analog circuitry, which must be carefully understood.

4. WRITE DRIVER CIRCUIT

In order to read data from an SRAM, the data must first be written into the memory. The circuit that writes the data into the cells is called a write driver or, from time reminiscent of only magnetic media, is called a write head. These terms are used interchangeably.

In typical high-performance SRAMs, as has already been discussed, a pair of differential bit lines is attached to each cell and these bit lines are pre-charged into a high state. The cells have transfer devices that are NFETs. Thus the transfer devices drive a strong "0" but do not drive a "1" very effectively. The writing a cell is accomplished by writing a "0" into either the true or the complement side of the cell and the cell latch causes the opposite side to go to a "1" state. It can be viewed that a cell passes only a "0" and never a "1". Since the transfer device passes only a "0" the write head need only drive a "0". A simple write head is shown in Figure 2-21.

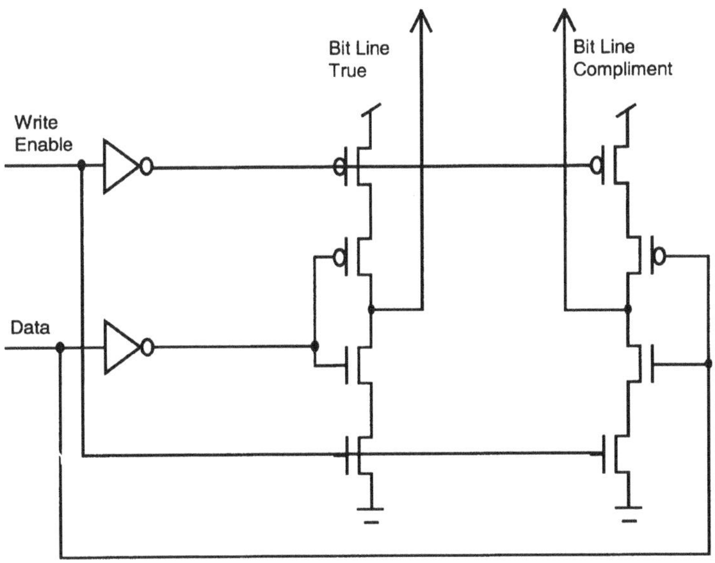

Figure 2-21. Write driver circuit.

The four vertical devices in series are often referred to as a gated inverter. When write enable (WE) is asserted high the write driver circuit drives the appropriate data values onto the bit line true and complement lines. Since the primary objective is to drive a "0", the NFETs and PFETs may be similarly sized, rather than the typical two to one ratio for PFETs to NFETs, as is used in most logic. Another alternative for write driver circuitry is to simply have one side driven through a pull-down device pair with the opposing bit line tri-stated in the high state. Further, the bit line pre-charge devices may remain on during a write, so long as the write driver circuits are strong enough to drive a solid "0" on the line that is needed. If this is used then the bit line equalization or balancing PFET is normally omitted from the pre-charge circuit or selectively turned off during a write.

5. DECODER CIRCUITRY

An address arrives at an SRAM boundary to identify which address is being selected. Since the RA in SRAM stands for "random access" any location may be selected on any given cycle. The same address may be selected over and over again, as in the case of a control store memory; sequential addresses may be selected, as in the case of an instruction cache. Lastly, various locations may be accessed in no particular order, as in the case of a data cache.

To identify the specific location, a group of address bits are supplied to the memory. From a user's perspective, the type of addresses, i.e. row and column, are non-consequential. The customer only wishes to be able to retrieve the data they stored at some time in the past. Functionally, the customer may rearrange the address signals in any order. The only concern is that the location selected for a given data input match the address location selected to obtain that same data on the output.

From a designer's or a test engineer's perspective the specific address type is paramount in consideration. Typically there is row, column, and bank addressing. The row can be thought of as the "y" address dimension while the column may be considered the "x" address dimension. The bank address may be referred to as sub-array, quadrant, or some other term and may be considered a third dimension. Practically, there are only two physical dimensions in memories since wafers are planar. It is possible to stack bits on top of bits to some extent but broad use of a true third dimension will not be done on a large scale for a single chip for some time yet. Stacking of chips can give a true third dimension but that is not addressed in this writing, as the design and test implications deal more significantly with packaging issues rather than memory operation.

From this discussion it can be seen that bank addressing is really column addressing, carried out in such a fashion that the columns are spaced even farther apart. Thus, the design and test implications for bank addressing are very similar to that of column addressing.

Memories must have at least one dimension of addressing, and some smaller memories do have only row addressing. Often a memory will have row addressing to access 256, 512, or 1024 cells along a bit line. It is not required that a memory have a power-of-two address space, but most often it is convenient to have one that is. Having a non-power-of-two address space means that there are invalid addresses, i.e. a specific address signal arrangement that does not go to any location. When non-power-of-two addressing is employed protection must be built in at the system level to prevent invalid addresses from being generated.

Column addressing typically selects between four, eight, or 16 bit-line pairs. Column addressing, however, is not necessary on some smaller memories.

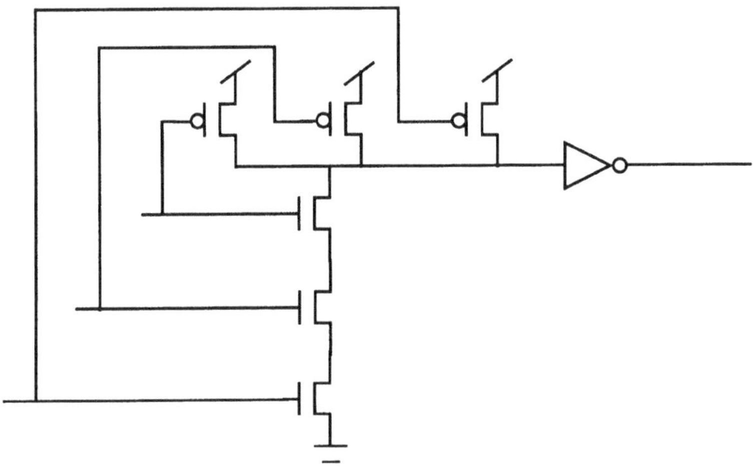

Figure 2-22. Example static decoder.

Decoders can employ either static or dynamic circuitry [23]. A static decoder looks like the three input static AND gate shown in Figure 2-22. A dynamic decoder requires fewer transistors since a full pull-up path is not required. Figure 2-23 shows a dynamic decoder, which selects from four possible address locations. Often times both types of decoders can be seen on a single memory since it is possible to utilize one type of decoder for row addressing and another type of decoding for column addressing. Also, the first and second stages of decoding can vary, where one uses static while the other uses dynamic. There are design trade offs for each when considering power, glitch-free operation, and possible faults. If a static decoder is utilized, often the last stage is of the clocked-static variety so that multiple addresses are not allowed to be addressed simultaneously, even if only momentarily.

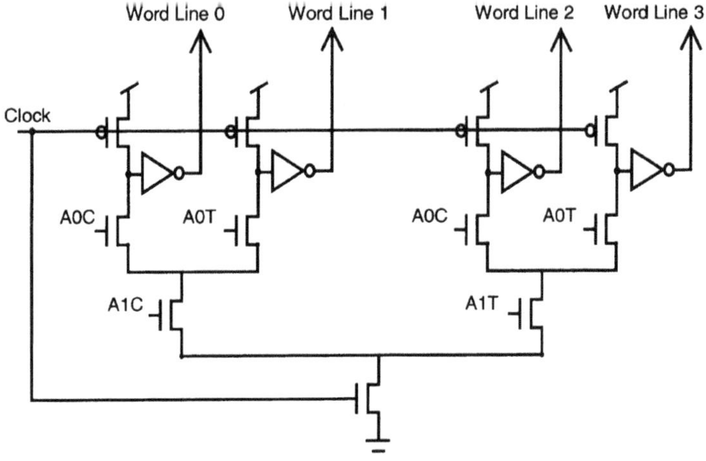

Figure 2-23. Example dynamic decoder.

6. LAYOUT CONSIDERATIONS

Earlier in this chapter the cell layout was considered. By examining Figure 2-7 it can be seen that there are certain symmetries and certain asymmetries within a cell layout. Optimizing the overall memory size can be accomplished by performing cell stepping, mirroring, and rotating of the cell layout. This stepping, mirroring, and rotating can also be done on a sub-array basis and full appreciation of each is required to adequately test the memories.

The cell of Figure 2-7 is repeated in Figure 2-24 but with the true portions of the cell highlighted. By removing the complement portions of the cell the stepping, mirroring, and rotating can be more easily explained. Figure 2-25 shows the true portion of the cell from Figure 2-24 stepped horizontally and vertically so that four cells are now represented.

Normally the cell is mirrored about the x-axis to facilitate the use of a single N-well that encompasses the PFETs from two vertically adjacent cells. Such a configuration is shown in Figure 2-26. The cells are simply stepped in the x dimension.

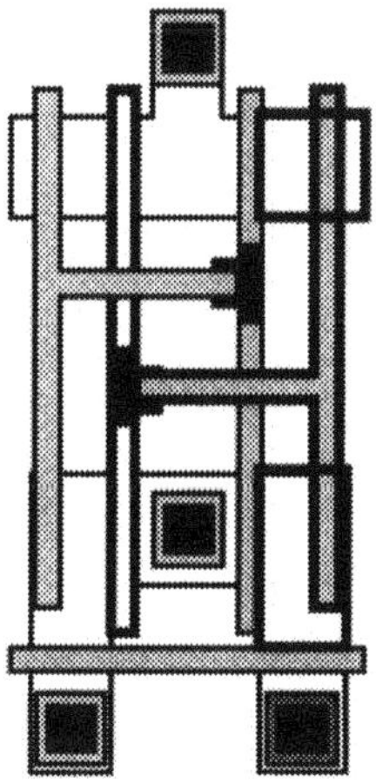

Figure 2-24. SRAM cell layout with true nodes emphasized.

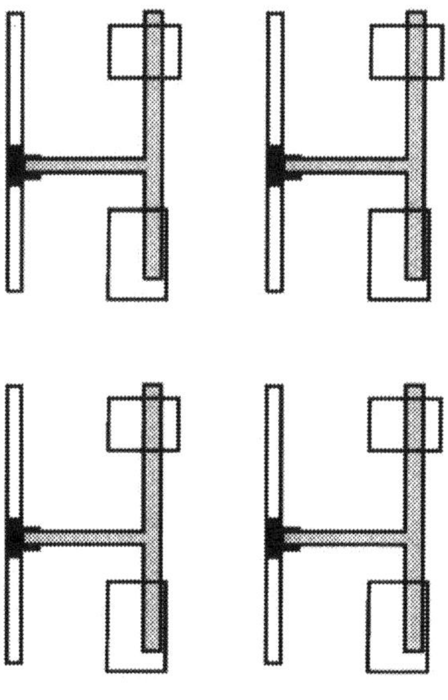

Figure 2-25. Layout from four stepped cells.

Often times the cells are mirrored about the y-axis as well to facilitate bit line layout. This is shown in Figure 2-27. It should be noted that just because the layout has been mirrored about the y-axis does not mean that the bit line true and complement signals have been swapped as well. Since a

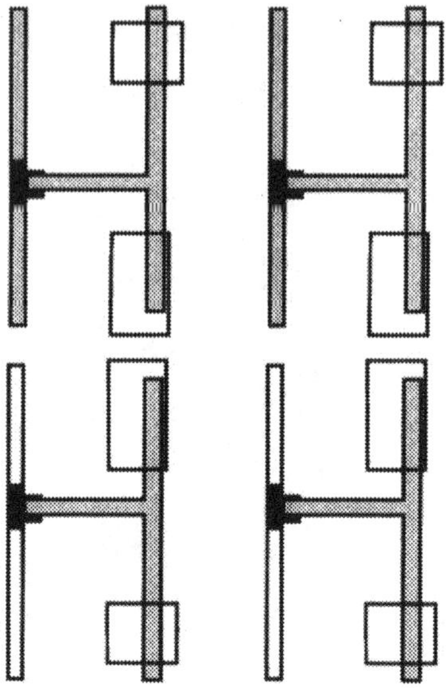

Figure 2-26. Four cells mirrored about the x-.

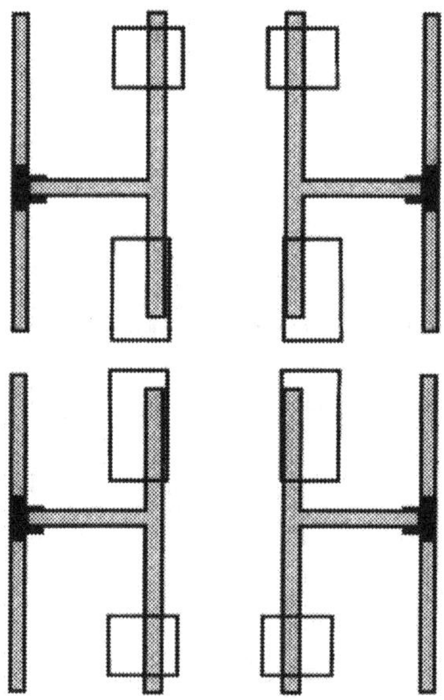

Figure 2-27. Four cells mirrored about the x-axis and the y-axis

cell is schematically symmetric the true and complement nodes can be either left or right without regard for relative stepping and mirroring.

All of today's memories have more than a single I/O, and most have very many inputs and outputs [24,25]. It is not unusual to have 256 or more I/O on a given memory. The multiple inputs are stored as a "word" in the memory. Most people assume that cells storing the bits from a single word are adjacent to one another, but they are not. From a logical point of view this makes sense but physical arrangements do not reflect this logic. In fact, the bits within a word are spread across multiple sub-arrays [26]. Only in certain very low power applications will the cells for a word be adjacent to one another. Figure 2-28 illustrates a four-bit word being within a sub-array (a) or being spread across four sub-arrays (b). For the latter case, the cells storing bit 0 through bit 3 are in row 127 and column 0 of sub-array 0 through sub-array 3, in this example. It can be seen that numerous cells exist between adjoining bits in a single word. One of the reasons for this arrangement is protection against soft errors. Because the cells within a word are not adjacent, if two physically adjacent bits flip due to a soft error, each word that is read will be detected as being erroneous since each contains a parity fail. If two bits in a single word were allowed to flip then no parity error would be detected.

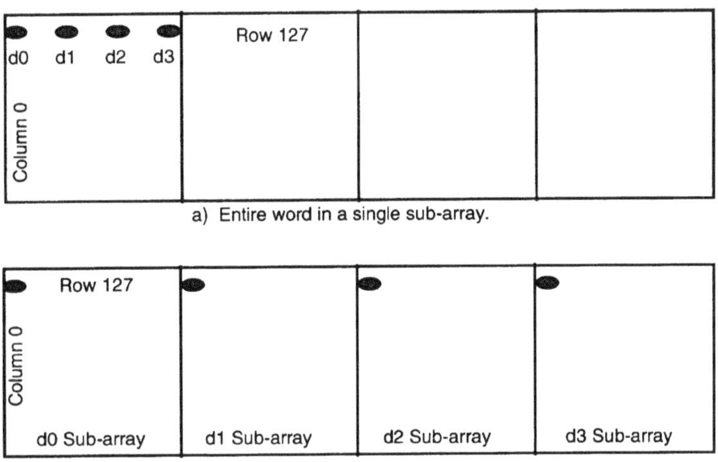

a) Entire word in a single sub-array.

b) Word spread across multiple sub-arrays.

Figure 2-28. A four-bit word spread across single and multiple sub-arrays.

Another physical layout arrangement involves bit line twisting. A pair of bit lines runs vertical defining a column of cells. The bit lines are long and are heavily capacitive. There is also significant capacitance from a bit line to the neighbor. Because of this significant line-to-line coupling capacitance between bit lines, to reduce having a single bit line couple into an adjacent bit line by too much of an extent, the bit lines are twisted [27]. Figure 2-29 shows a typical set of twisted bit lines. One pair will be twisted at the half way point. The next pair will be twisted at the one quarter and three quarter points. This arrangement is referred to as a triple twist [28]. Only one quarter of any given bit line will be coupled into its neighbor. Additionally, due to the location of the twists, one quarter of a bit line's capacitance will be coupled into its neighbor pair's true line and one quarter will be coupled into its neighbor pair's complement line. This arrangement reduces the amount of coupling into any given bit line and forces any coupling to be common mode, on the true and complement bit lines. Since the coupling into the true and complement nodes is common mode, the sense amplifier is far less prone to erroneous evaluations. Other possible bit line twisting arrangements exist but this one is the most common.

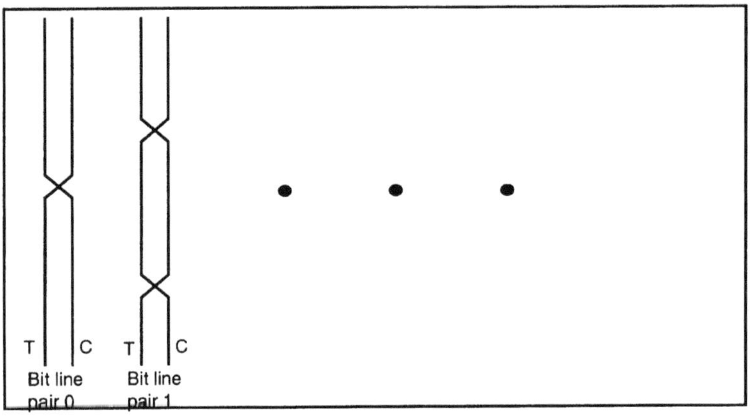

Figure 2-29. Typical bit line twisting arrangements.

7. REDUNDANCY

Memories require redundancy to ensure that sufficient chip yield is obtained. A redundant element is a piece of memory that can replace a defective piece of memory. Redundancy can come in the form of spare rows, I/O, columns, blocks, or a combination of the above. Very small memories can get by without redundancy but large memories require significant numbers of redundant elements. When there are many small memories on a single chip, again somc form of rcdundancy should bc included or else yield will be negatively impacted. As process technology improves, the amount of memory that can be included without the need for redundancy increases. When a new smaller lithography is achieved, there is a greater need for redundancy until that process matures. Even as one looks across generations of technology, the need for redundancy decreases on a per bit basis. Since the number of total bits per chip is growing at a very substantial rate, however, the total amount of redundancy needed is growing on a per chip basis. At one point redundancy was typically included on SRAMS of 64 Kb or greater [29]. As of this printing, most producers are including redundancy when memories get larger than three quarters of a megabit to 1.5 megabits [30]. This trend will continue, but ever larger numbers of cells are being packed on chips. Thus, more memory with more redundancy will be included as time continues.

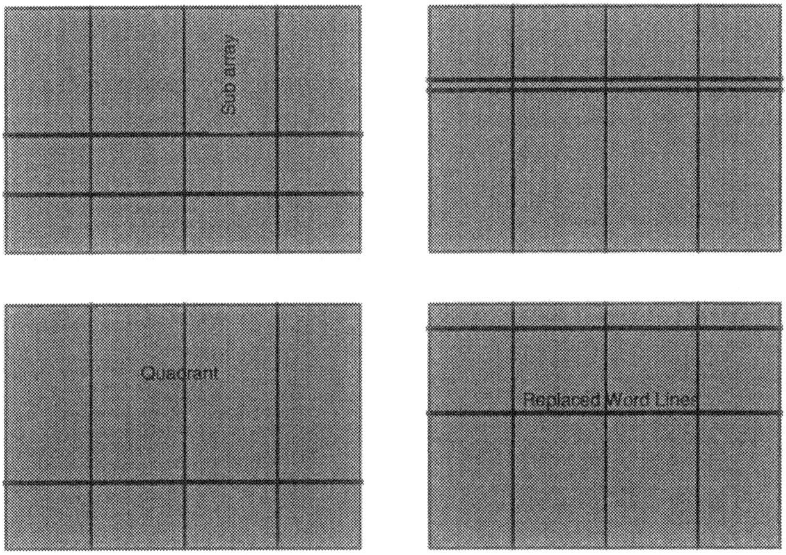

Figure 2-30. Independent redundancy replacement illustrated in four quadrants.

Row redundancy is implemented by including one or more spare word lines in a memory. This redundancy is utilized when a bad cell, row, or partial row needs replacing. A set of latches or fuses determines which row address is to be replaced, with a comparator in the memory's decoder examining these latches. When an address is functionally supplied to the memory, if a match to the pre-defined value in the latches or fuses is detected, the redundant row is accessed and the normal word line, which is defective, is not accessed.

When a spare I/O is included in a memory design there are usually a large number of I/O in the memory. For example, if the memory has 64 functional I/O, a spare or 65th I/O is included for redundancy. I/O replacement is a form of column redundancy and is easy to implement. True column redundancy replaces one or more bit-line pairs in the memory. For this type of redundancy, a column is replaced within an I/O to fix a failing cell, sense amplifier, pre-charge circuit, bit line, or partial bit line. Block redundancy is utilized to replace a larger portion of memory. In this case rows, columns, and I/O are all replaced.

Figure 2-30 shows an example of redundancy replacement in a memory. It can be seen that each quadrant has independently controllable redundancy. In the upper right quadrant a row pair has been replaced. In the upper left quadrant two independent rows have been replaced. In the lower left quadrant only a single row has been replaced, meaning that one spare row was not utilized in this quadrant. In the lower right quadrant, again two independent rows have been replaced. More information on redundancy will be covered later in the text in chapter 13 on BIST and Redundancy.

8. SUMMARY

Static random access memories are the primary form of embedded memories in the industry today. They have great utility and robustness. There are, however, many subtle analog effects that need be considered from both a design and a test perspective.

The sensing scheme, decoder circuitry, redundancy, and layout arrangements all bear on the other memories, which will be covered in the remainder of this text. Therefore, it is recommended a finger be kept in this chapter as one reads of the other memory designs. In multiple places the reader should refer back to the SRAM circuitry for detailed explanation. This has allowed the overall text to be shorter and the writing to be less pedantic.

SRAMs are easier to comprehend than most other memories because of their closeness to logic and therefore were covered first. Unique complexities due to technology, embedded logic, or other challenges will pervade the memories to be discussed in the next chapters. Thus, the SRAM is the standard in addition to being the workhorse of the industry. Comprehending the design and test challenges from an SRAM perspective will provide invaluable assistance in understanding other memories.

Chapter 3

Multi-Port Memories

Design & Test Considerations

"He has told us that you always have pleasant memories...." – the apostle Paul in I Thessalonians 3:6

Multi-port memories have all the features of single-port memories, only more so. When you think of multi-tasking, the multi-port memory is a true example. Multiple simultaneous operations go on in parallel. Depending on the memory architecture, the operations can even be asynchronous. The design complexities and domain space for variations are huge. There are numerous applications where a multi-port memory is quite useful. The simplest case is a memory where a system engineer would like to write data each cycle and would also like to read data each cycle. This is the case in a branch-target-buffer memory in a microprocessor [31]. During each cycle, the buffer memory keeps track of branch-prediction results and uses those results to determine if the current instruction is probably a branch [32]. Keeping track of a branch-prediction result involves writing. Determining if the current instruction is predicted to be a branch involves reading. Thus, a read and a write go on simultaneously each cycle.

Another example where a multi-port memory is utilized is in a networking environment. In this case a slew of operations are going on simultaneously and there can be numerous requests for data from various operations. A third example would be a multiprocessor environment where more than one processor can be calling for data from a memory. The examples can go on and on.

Our minds can have only a single cognitive thought at a time. As an example, an author cannot be typing a book with the right hand while at the same time typing an article with the left hand. There can be numerous things going on "in the back of our minds" and there are automatic functions such as breathing, circulation, and so on that are maintained while our mind is occupied elsewhere. A multi-port memory allows two "cognitive" exercises to be performed simultaneously. It is like having two people looking at two different books in one library at the same time. The library provides multi-port operation. The challenge comes when two people want to use the same book at the same time. A conflict occurs in this case and a similar conflict can occur in multi-port memory addressing as well.

The number of semiconductor multi-port memory applications will continue to grow as system designers strive to increase performance and recognize the availability and utility of multi-port memories. This realization will cause greater use and demand for increased complexity in multi-port memories. The key place that this complexity is seen is in higher dimension multi-port memories. It is not unusual to see two, four, six, or nine port memories. Even higher dimension multi-port memories can be encountered, especially when considering pseudo multi-port memories, which will be discussed later.

1. CELL BASICS

A multi-port memory cell has to have more than one access port. At a conceptual level Figure 3-1 illustrates the single port memory cell and the multi-port memory cell, where there is more than a single bit line contacting each cell along a column.

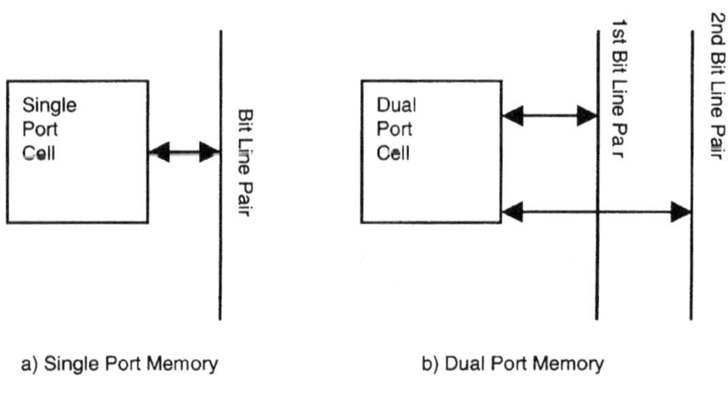

Figure 3-1. Block diagram of a multi-port memory column cell.

For this section of the book we will assume that a multi-port memory is an SRAM. It is possible to have other types of multi-port memories but SRAMs are the most predominate variety and should remain so. Further, the challenges seen in other types of multi-port memories are similar to that seen in SRAMs. Simply for reference, Figure 3-2 shows a dynamic multi-port memory cell [33].

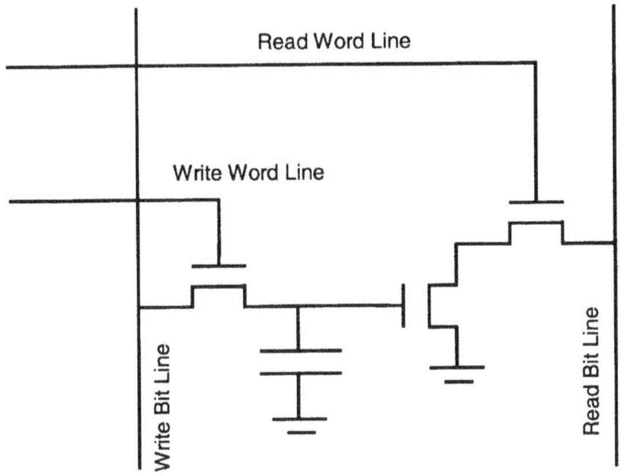

Figure 3-2. Example DRAM multi-port memory cell.

An example two-port multi-port memory cell is shown, at the schematic level, in Figure 3-3. This is the simplest of two-port memory topologies. There is a single write bit line and a single read bit line, as well as a single transfer NFET writing the cell and a single transfer NFET for reading the cell. Since a single ended read approach is utilized, a read on the bit line must utilize a large swing to correctly evaluate the cell's contents. The read bit line can be pre-charged to a high value. When the cell is read, the bit line discharges toward ground. The sensing scheme can be simple in that an inverter can suffice at the output of the read bit line. A read operation will be relatively slow with such a topology.

Writing this type of simple two-port memory cell is a delicate operation. An NFET transfer device is very effective at writing a "0" but very ineffective at writing a "1". Due to this fact, the write NFET transfer device needs to be larger than normal to be able to adequately write the cell to a "1" state. Writing to a "1" is again a slow operation due to the limited pull-up capabilities of the transfer NFET and time must be allotted to ensure that the "1" made its way into the cell. Generally, this topology cell should be avoided.

While the previous example of a two-port memory cell makes explanation of the multi-port memory concept easy, it is rarely used due to the stated read and write speed limitations. Figure 3-4 shows the most used multi-port memory, the standard two-read/write port cell [34]. This two-port memory cell utilizes a differential read and a differential write. It has eight devices and is very similar to the six-device single port memory cell, which has already been discussed. The only additions are the two transfer devices for the second port, shown in Figure 3-4 as T7 and T8. These sink the appropriate charge from the corresponding bit line during a read operation. During a write operation the desired cell nodc is pulled low. It can be seen that an extra bit-line pair and an extra word line must be attached to the cell for the additional port. When this type cell is used for two read/write ports, it is often referred to as a dual port memory. When this cell is utilized with one port being a dedicated read port and the other as a dedicated write port, it is often referred to as a two-port memory.

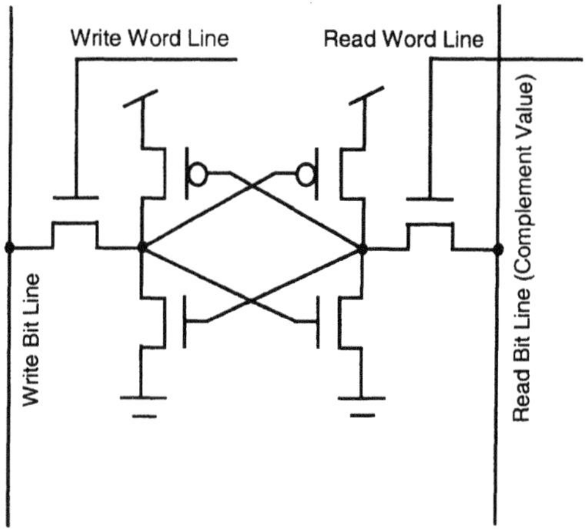

Figure 3-3. Simple two-port cell with single-ended read bit line.

From this schematic it is obvious that there are two word lines, one for port 0 and one for port 1. Further there are two bit-line pairs. One pair is the true and complement bit lines for port 0 and the other is the true and complement bit-line pair for port 1. From the shear number of lines which must interact with and therefore intersect the cell's x-y location, it can be seen that this two-port cell takes up significantly more room than a one-port cell. Generally an eight-device cell takes up twice the area of a typical six-device cell. The extra space is driven by the extra bit and word lines rather than the active shapes.

The stability of a multi-port cell is much trickier than a one-port cell. It must be realized that, since two simultaneous operations are performed regularly, it is possible to have both ports reading the same location at the same time. As was addressed in chapter 2, read stability depends on the relative strength of the pull-down and transfer device. In the case of a multi-port memory cell it is the relative strength of the pull-down device to all of the transfer devices combined.

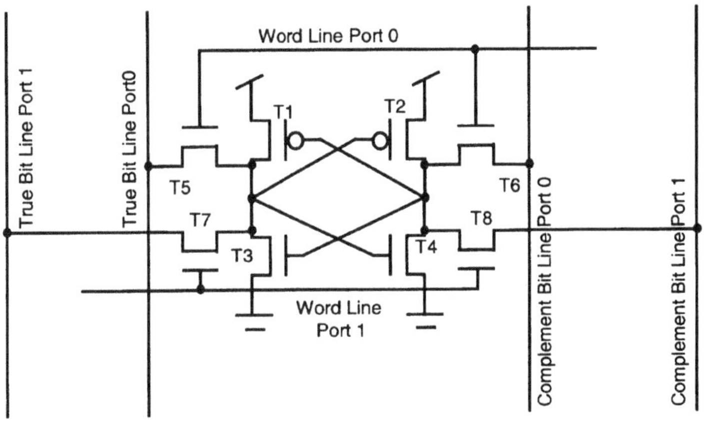

Figure 3-4. The eight-device two-port read/write cell.

The extensive complexity of multi-port memories is shown when more than a single port is operated simultaneously for a single cell. If two reads are performed on a single cell then the cell has to be able to sink twice the current of a single ported memory. The beta ratio, which is so critical to single-port cell stability, is now the strength of the pull-down device divided by the strength of both transfer devices for a two-port memory. Equation 3-1 shows the beta ratio for an "n" read port memory.

$$\beta = \left(\frac{W_{eff}\,PullDownFET \;/\; L_{eff}\,PullDownFET}{\sum_{n=1}^{Number\_of\_Read\_ports} (W_{eff}\,TransferFET_n \;/\; L_{eff}\,TransferFET_n)} \right)$$

Equation 3-1. Beta ratio calculation for an "n" read port memory.

Achieving a high enough beta ratio, typically around 2.0, is a requirement to maintain cell stability. It is critical, though, that the cell be writable by a single port. A cell can be designed that is so "stable" that it cannot be written. Thus, when a high dimension of read ports is desired, an active-read capability can be provided as shown in Figure 3-5. Here transistor T9 actively pulls the bit line low when word line 0 goes high and a "0" is stored in the cell. No concern about beta ratio is required since the cell is not required to sink the current from the bit line. This type cell cannot be disturbed by a read operation. In this cell there is one write port and two read ports. Each read operation is performed differentially and the only load on the cell during a read is the capacitance of the gates from the active pull down NFETs.

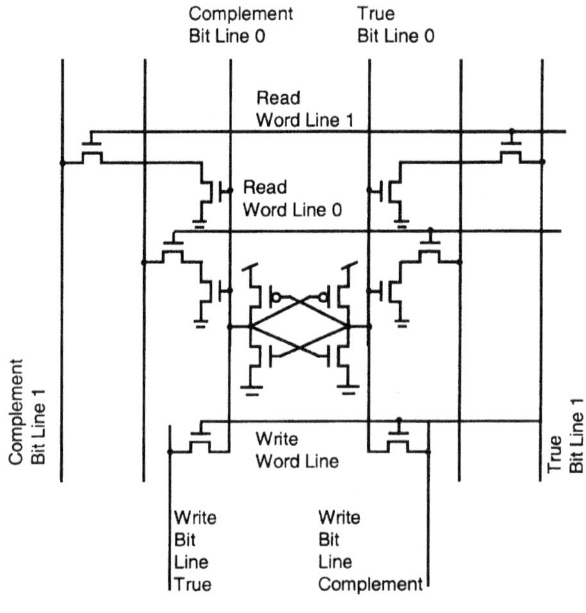

Figure 3-5. Active pull down on cell with three read ports.

It is also possible to utilize an active pull-down NFET on a single-ended read. The pull-down devices can be larger than typical transfer devices, since the memory designer does not need to be concerned with disturbing the cell on a read. The large pull-down devices allow a faster bit line discharge rate and therefore a faster read operation. This means that a full swing or a virtually full swing is possible on a read. For a differential read it is also possible to leave the pre-charge devices on, since the larger active pull-down device can overcome the pre-charge FETs and generate the required small differential signal. If an equalization PFET exists between the differential bit lines it should be turned off during a read so as not to discharge the unintended bit line and thereby reduce the differential signal.

Multi-port memory cells can become quite complex since there can be very many ports which access each cell. An example five-port cell is seen in Figure 3-6 [35]. This cell utilizes one single ended write port and four differential active pull-down read ports.

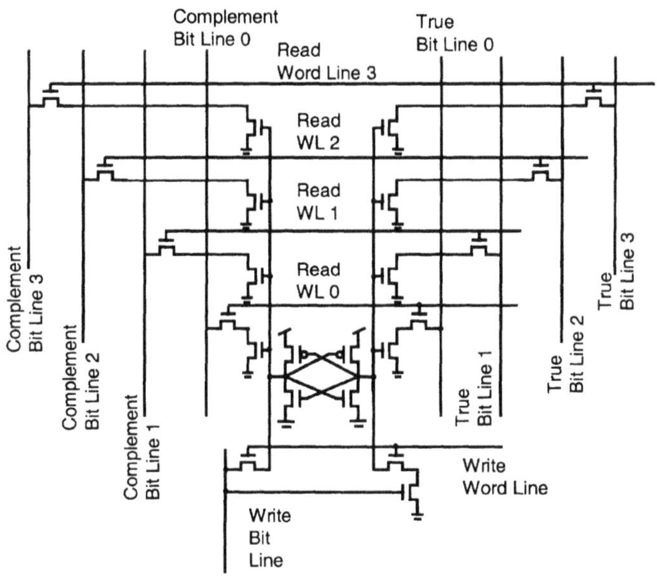

Figure 3-6. Five-port cell with four active pull-down differential read ports.

2. MULTI-PORT MEMORY TIMING ISSUES

Memories can be true multi-port memories or they can be pseudo multi-port memories. A pseudo multi-port memory utilizes a high-speed clock to generate multiple memory accesses in a single system cycle [36]. This is sometimes referred to as time division multiplexing (TDM) since two time slices or divisions exist to access or multiplex into a given memory location. The two clock pulses are normally generated within the memory and utilize the memory's own timing chain. When the first batch of data is locked into the memory's output latches or when the first write operation is completed, the second memory operation is started. In this manner two, or even more, memory accesses can be designed into each system cycle. A TDM operation can be performed on a memory that already has a multi-port cell. In this manner a higher dimension pseudo multi-port memory can be designed. The five-port cell in Figure 3-6 is implemented so that the read ports are double clocked but the write port is only clocked once per cycle. This allows a five-port memory to provide a pseudo nine-port memory operation. The area of the five-port memory cell is much less than a similarly designed nine-port cell. The key requirement of a nine-port memory is that the memory needs the ability to read at twice the system cycle rate.

A true multi-port memory can be either synchronous or asynchronous. In the case of a synchronous two-port memory, there is only a single clock. This clock fires both ports' operations at the same instance in time, thus both word lines go active at the same time and the design is well controlled. An asynchronous two-port memory has two clocks. One port's operation can happen at the same time or at entirely different times from the other port. There may indeed be overlap in the timing between the two ports and because of this there are certain legal and certain illegal operations and timings that are allowed. For instance, if both ports are accessing the same location and one of them is writing while the other is reading, legal possibilities need definition. If the write clearly occurs prior to the read then the new data is read. If the write clearly takes place after the read then the old data will be read. In between these two timing extremes, the data being read is indeterminate and needs to be specified as not being allowed or specified as having the outputs be Xs. One last thought on writing and reading the same location is that a read creates a load on the cell. Writing a cell that is being read requires driving a greater load than writing a cell that is not being read. Therefore, a multi-port memory write driver needs to be designed with this anticipated greater load and greater drive capability in mind. The write time for an individual cell is longer as well and so the word line up time and entire timing chain needs to factor these timings into the design.

3. LAYOUT CONSIDERATIONS

Multi-port memory cells are significantly larger than single-port memory cells because of the higher the number of ports. As the number of ports grows so does the number of word lines and bit lines that contact the cell. The multiplicity of operations that is performed creates interaction coupling challenges in a design. A read operation causes a small amount of signal to be developed over time on a bit line which is, in turn, sensed by a sense amplifier. A write operation causes a bit line to rapidly transition from Vdd to ground. Any line that is adjacent to the write bit line, which is driven low, will have current coupled into it according to Cdv/dt. If a read bit line which is supposed to stay high is immediately adjacent to this write bit line, it will be coupled down thereby reducing the signal that is propagated to the sense amplifier. The read bit line is only held high capacitively. Having the read bit line coupled low can cause catastrophic problems. Therefore, shielding is frequently designed into the multi-port memory to prevent excessive coupling into the read bit lines. The shielding lines can be alternating ground and Vdd lines as shown in Figure 3-7 and can also be similarly used in the word dimension.

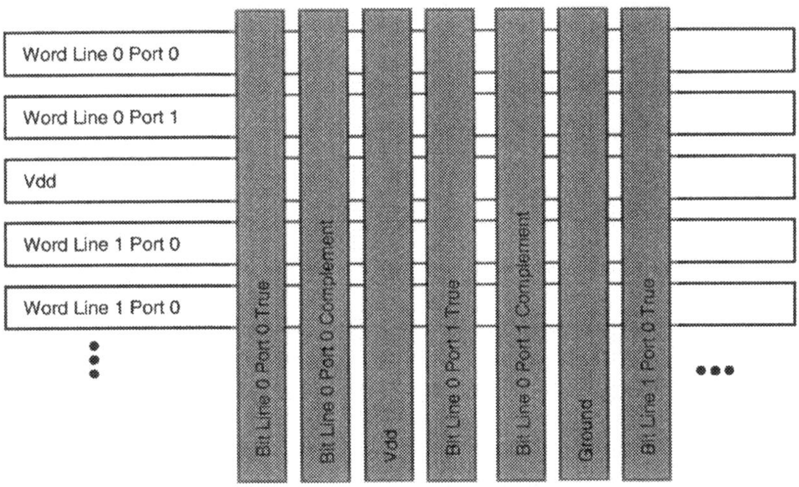

Figure 3-7. Two-port memory bit lines with shielding.

A multi-port memory cell often is asymmetric. While a one-port cell can easily be flipped in the x and/or y dimensions, a multi-port cell normally cannot. This is especially true when the number of ports gets larger and when there is an odd number of those ports. Therefore, it is often only possible to step a multi-port cell and not mirror, flip, or rotate the cell.

Key Point: Multi-port memory designs must consider all of the possible interactions between the ports.

Since there are so many bit-line pairs on a multi-port memory, bit line twisting can become very difficult. On higher dimension multi-port memory cells the true bit lines may be grouped together vertically and the complement bit lines may be grouped together in another location vertically. When this happens it is impractical to perform bit line twisting. With no bit line twisting the coupling concerns need to be examined more carefully with detailed simulation of possible noise coupling to ensure the minimum amount of bit line signal development happens. This ensures repeatable and reliable cell data evaluation. As multi-port memories become larger and more prolific, the concern only becomes larger. This complexity needs to be factored into future multi-port memory design efforts.

As multi-port memories become larger, more and more of them will need to include redundancy to enhance yield. Clearly, if one port on a cell fails then both ports need replacement. One cannot write into one working port on a normal element and then read from a redundant element on a different port and expect the data written into the normal element to be present. Thus a failure, on any port, must cause complete replacement of the bad element with a redundant one.

4. SUMMARY

Multi-port memories have many similarities to standard one-port SRAMs. Due to the multiplicity of ports, the number of possible interactions between the ports creates significant complexity. The multi-port complexity must be simulated from a design perspective and analyzed from a test perspective. Inadequate design analysis can easily result in an un-writable cell, one that disturbs if multiple ports read the same cell, or coupling between the write and read ports. Inadequate analysis from a test perspective will not consider all of the possible multi-port faults resulting in poor test patterns and higher shipped defective chip rates.

Chapter 4

Silicon On Insulator Memories

Design & Test Considerations

"...was to him but a memory of loveliness in far days and of his first grief...." – from *Return of the King,* J.R.R. Tolkien

Silicon on insulator or SOI memories have come to the forefront in the last few years, especially in high performance applications [37]. Silicon-on-insulator technology has existed for a considerable period of time but only reached widespread use recently. SOI provides significant performance advantages, sometimes as much as 20% or more [38], due to technology improvements. These improvements are reviewed here along with the resulting increase in design and test complexity. SOI technology is the next logical step, as this book works its way outward from the base SRAM into memories and technology that progressively becomes more varied. Variations from the base SRAM drives complexity from a test and design perspective that clearly needs understanding.

1. SILICON ON INSULATOR TECHNOLOGY

Bulk silicon has been the standard processing technique utilized in most semiconductors. Silicon on insulator provides a marked departure from bulk CMOS [39]. Figure 4-1 shows a typical bulk transistor cross section. The gate, source, and drain are each directly contacted to provided the needed switching operation. The substrate is seen at the bottom of the figure and is electrically continuous down through the silicon. In the case of an NFET, the source and drain are doped to be n+ silicon while the substrate is p type. In the case of a PFET, the source and drain are p+ silicon while the substrate is n type. Either can be in a well or tub, depending on the chosen process.

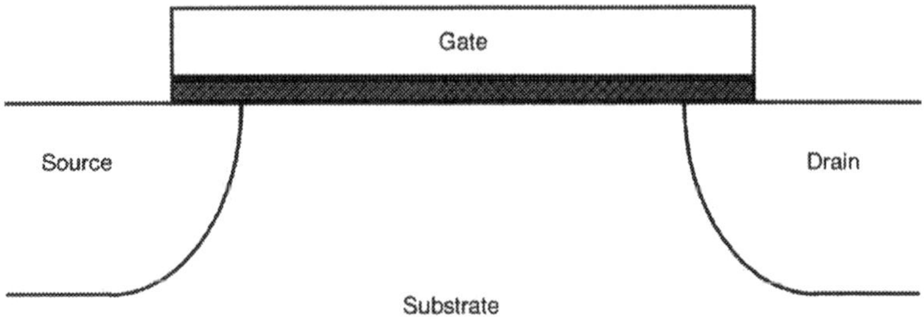

Figure 4-1. Bulk silicon transistor.

For a silicon-on-insulator transistor the source, drain, and gate structures are directly contacted. The fourth connection or body, though, is interrupted by a buried layer of oxide [40]. The oxide forms the insulator on which the rest of the transistor sits, as shown in Figure 4-2. A cross-section photograph of two SOI transistors and the supporting metallurgy is shown in Figure 4-3. Since the oxide is underneath the source and drain, the junctions are reversed biased and the depletion region cannot grow as large as that of bulk silicon. Because the depletion region is smaller, less charge can be stored and thus there is less diffusion capacitance. This reduced capacitance is one of the largest contributors to the improved performance in silicon-on-insulator technology.

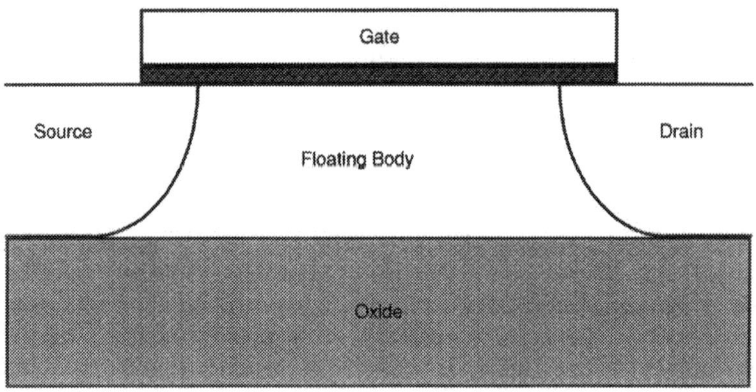

Figure 4-2. Silicon-on-insulator transistor.

The other performance contributor in SOI transistors results from the floating body. Since the body of the FET is not electrically tied to a common substrate layer in the silicon, its potential is allowed to float. There are parasitic diodes between the body and the source/drain which limit the voltage range over which the body potential can vary. There are also capacitances between the diffusions and the body as well as between the gate and the body. The parasitic diodes and capacitors are shown in Figure 4-4.

Figure 4-3. Cross-section photograph of SOI transistors and metallurgy (photo courtesy of IBM Corporation / Tom Way Photographer).

Since the body is allowed to vary in potential, the threshold voltage of the FET can change with it. For an NFET, as the body potential rises the threshold voltage drops. This results in a transistor that turns on earlier and has more available overdrive. Equation 4-1 shows the relationship between the body voltage and threshold voltage [41]. VT0 is the base threshold voltage; V_{SB} is the body to source potential. The symbol γ is the body-effect constant and ϕ_F is the Fermi potential.

$$V_T = V_{T0} + \gamma\left(\sqrt{V_{SB} + |2\phi_F|} - \sqrt{|2\phi_F|}\right)$$

Equation 4-1. Threshold voltage as a function of body potential.

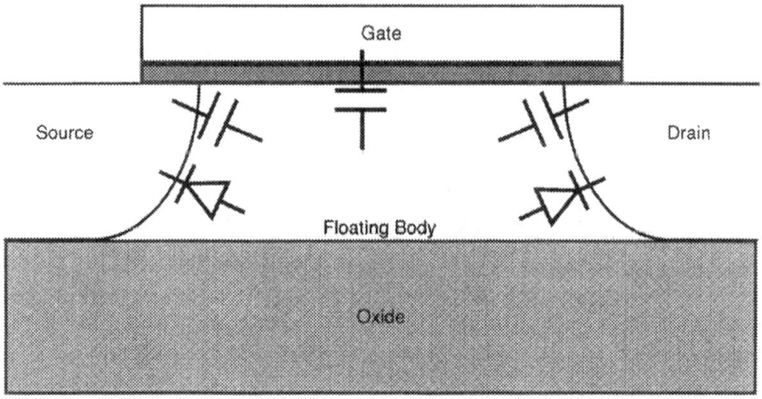

Figure 4-4. Silicon on insulator NFET with parasitic diodes and capacitors.

As stated earlier, as the body potential varies so does the performance of the transistor. The body potential is a function of the last operation performed on the FET and the time since that last operation. The variability of the FET performance is known as the history effect. For an NFET, if the drain is high and the body has been coupled up, the body's potential could be as high as a diode voltage drop above ground, where the source is attached to ground. The threshold voltage is then at its minimum and the drive capability is at its maximum. When the gate is transitioned high, the NFET turns on very quickly and rapidly pulls the drain down in potential, coupling the body down as well. If the body is low in potential, even below ground, then the drive capability is diminished, as is the case after the first high speed switch. The method of silicon-on-insulator fabrication can generate two different types of depletion and these two have different results. Fully depleted silicon on insulator is found more readily in thin active silicon layers and the history effect is minimal. Partially depleted silicon-on-insulator technology is more compatible with bulk device physics and processes, uses thicker structures, and can have a significant body effect [42]. Designers must factor this history effect and its impact of varying device performance into the chip timings.

2. MEMORIES IN SOI

A memory processed using silicon-on-insulator technology will have differences from those developed for standard bulk silicon [43]. SOI can be utilized to fabricate dynamic RAMs [44], but mostly it is used for fabricating SRAMs. Since the FET structure in SOI can have a varying threshold voltage, it can also have a varying sub-threshold leakage current. Dynamic RAMs need to limit their leakage current as much as possible to lengthen their data retention time. More information will be covered on dynamic RAMs in chapter six.

Silicon on insulator SRAMs are sensitive to certain design characteristics, which are not of concern in bulk SRAMs [45]. The cells along a column can behave in unusual manners due to the floating bodies of devices in the cell [46]. Taking the transistor with its parasitic diodes from Figure 4-4, it can be seen that a parasitic bipolar transistor exists, as shown in Figure 4-5. The body can form the base of a NPN bipolar, while the source forms the emitter and the drain forms the collector. If a body is high in potential and the source is pulled down, the parasitic NPN device may be turned on, pulling current from the body and also pulling current from the collector/drain due to the bipolar gain of the NPN device.

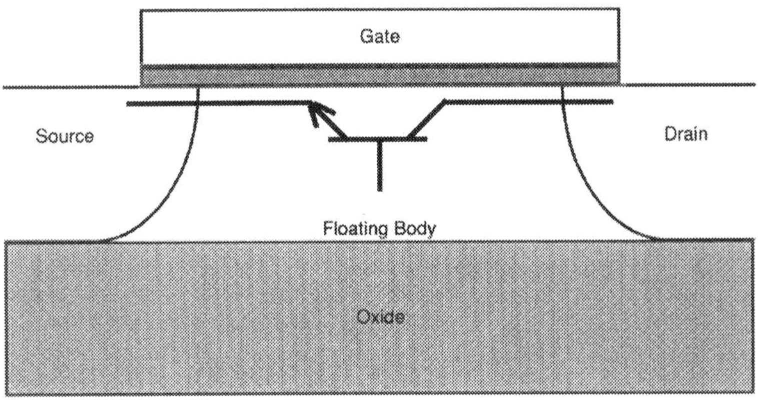

Figure 4-5. Parasitic NPN device in a SOI NFET.

The NFETs, which form the pass gates of an SRAM cell, can form just such a parasitic NPN where each cell is attached to a bit line. Figure 4-6 shows a series of cells along a column. The four transistors in each cell that make up the cross-coupled latch have been abstracted for simplicity. The transfer devices remain explicit and are shown along with their parasitic NPN transistors. Since the bit line is normally pre-charged to a high potential, the emitter is in a high state. If all the cells along the column are in a "1" state, the collectors are at a high potential as well. Since the emitter and the collector potentials of the pass transistors are high, their base potentials rise due to diode leakage currents. This state, with all of the cells along a column having a "1" value, may be unusual but it is a case which must be accommodated. Certainly during test, with typical marching and walking patterns, such a state will be experienced. When one cell is to be written to a "0" the write head for that column drives the bit line low. The load that the write head sees with a bulk silicon SRAM is the large capacitive load from the long line. With silicon-on-insulator technology, the load varies with the data stored in the column; the write drivers and cells need to be able to sink this variable load [47]. Due to the body potential's ability to rise towards the potential of the drain (bit line), with both at similar potentials, the space-charge region surrounding the drain diode becomes very thin, even thinner than bulk. This thinness essentially reduces the distance between the plates of the capacitor.

When the bit line is pulled low on an SOI SRAM, the parasitic emitters are all pulled low. When each emitter reaches a diode drop below the potential of the base, current starts to be pulled from the base. When current starts to flow from the base, the NPN transistor turns on and momentarily pulls current from the collector until the base empties of charge, according to Equation 4-2. The beta amplification factor is small but is a function of the layout, doping, and other specific technology values, which must be evaluated. Nonetheless, current will be pulled from the base and from the collector of each cell. The write driver circuit must be strong enough to drive the capacitive bit line low and be able to sink the bipolar NPN currents. Each of the cells must be stable enough so that when the current is pulled from the collector, also known as the cell's high node, the cell does not switch in value even with this bipolar gain.

$$I_E = I_B(\beta + 1)$$

Equation 4-2. Bipolar current equation.

If a full swing read is utilized on the bit lines, then each cell must be able to sink a similar current to that described in the previous paragraph. In addition, the cell which is sinking the current needs to be able to do so without causing any stability problems to itself. In Figure 4-6, the bottom cell is shown with a "0" on the true node. This cell would need sufficient drive to pull down the bit line and the associated parasitic bipolar current. Figure 4-7 provides another representation, with two cells along a column, emphasizing the parasitic NPN structures. For simplicity the transfer NFET is not shown, since it does not contribute to the bipolar current.

Key point: In SOI memories, the parasitic bipolar and history effects must be considered.

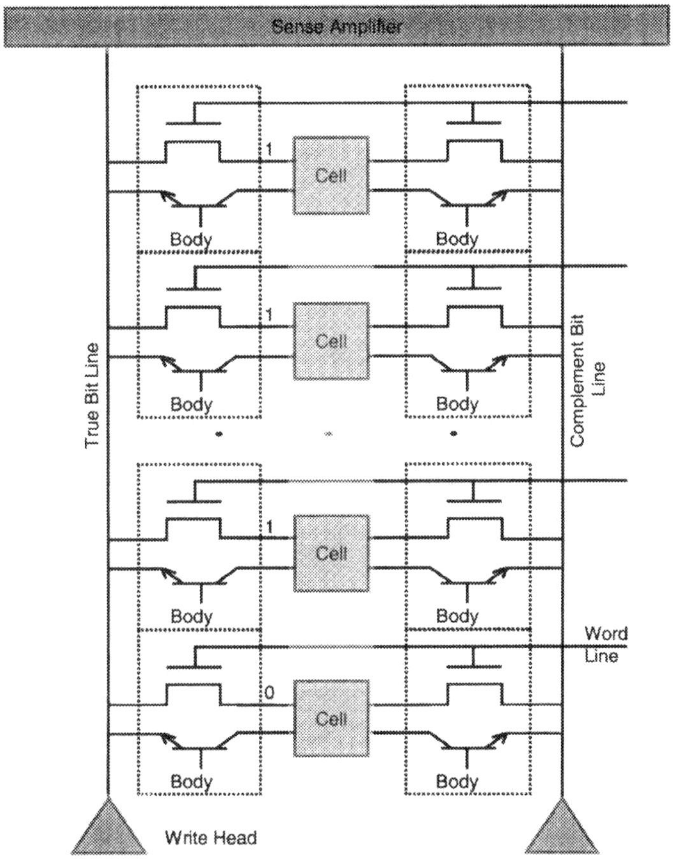

Figure 4-6. Column of cells showing the group of parallel NPNs.

Cell stability has some unusual factors in silicon-on-insulator technology. The cell stability, as stated earlier in chapter two, is governed by the cell's beta ratio. The beta ratio is defined as the ratio of strength of the pull-down FET to the transfer FET. In SOI, the strength of the FETs is a function of their most recent operation and the duration since that operation occurred. Thus the beta ratio changes with the recent history that the pull-down and transfer devices have been through [48]. As a result, the nominal beta ratio, based on the effective device widths and lengths must be higher than a typical bulk SRAM. This higher nominal beta ratio allows the cell to retain stability regardless of how the real FET strengths vary due to the history, with a small performance impact.

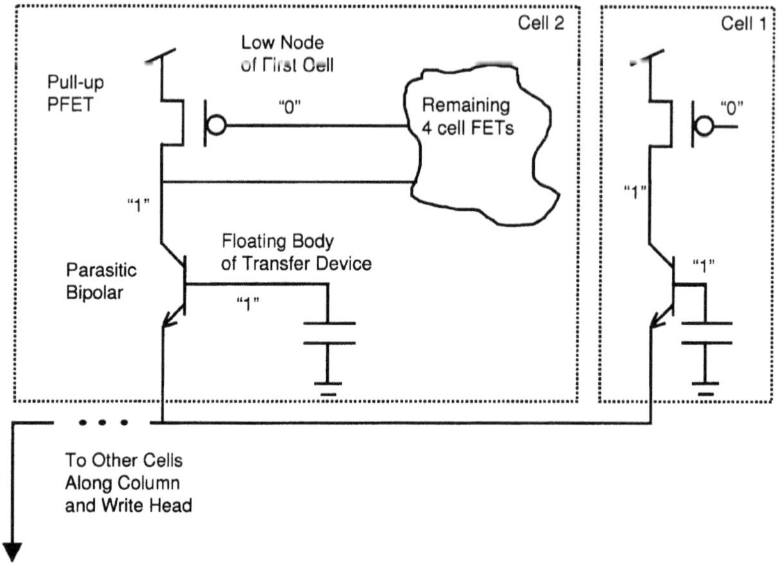

Figure 4-7. A cell with the parasitic NPN detailed in the context of a column.

3. LAYOUT CONSIDERATIONS

As just stated, the nominal cell beta ratio needs to be higher due to the history effect in silicon-on-insulator technology. Thus, the layout of the SRAM cell needs to be modified, as compared to bulk technology.

Other circuits are impacted by the history effect as well. Any circuit that is clocked regularly, and that works in tandem with another path, which is not clocked regularly, is suspect [49,50]. The circuit that is clocked regularly is often referred to as being in steady state while the circuit that is only clocked rarely is said to be in dynamic state. One example is a sense amplifier, which is clocked every few nanoseconds, and a particular cell, whose row may not have been clocked in hours. To compensate for this difference, more signal development time needs to be allocated to handle variances due to signal development that is a function of history.

Another concern with signal development is accumulating bias in the sense amplifier itself. If a sense amplifier has been reading one data type over and over again it can develop a bias for that data type. Then when the complement data polarity needs detection, the sense amplifier has bias that must be overcome. To prevent this from occurring, a body tie down can be provided. This layout feature allows a body to be contacted and biased to a specific potential. It also allows specific bodies to be tied together. Figure 4-8 shows a body contact point on an NFET, this being the fourth terminal which is normally not shown but which is assumed. In bulk CMOS the body of an NFET is tied to ground. In silicon on insulator the body is normally floating for best performance. In certain cases, as in the situation of a sense amplifier, it is desired to contact these bodies. Figure 4-9 shows an example sense amplifier schematic with the bodies tied on the primary sensing NFETs.

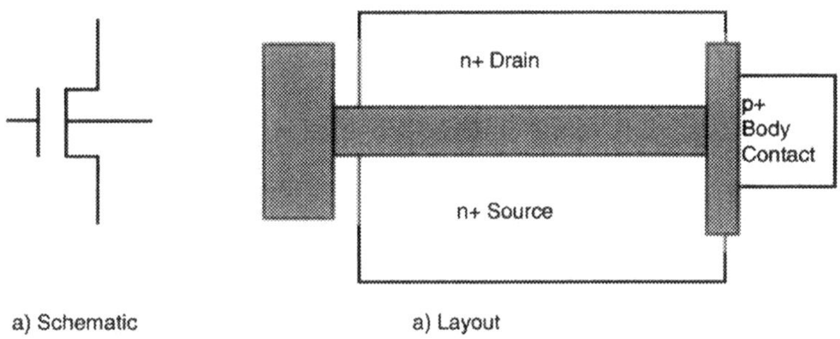

Figure 4-8. NFET body connection point, a) schematic and b) layout.

Body connections need to be used sparingly since they require significant area overhead and increase path delay. The body tie increases capacitive loading and decreases performance, thus making the circuit behave like bulk, or worse, and therefore not having speed performance advantages. A body contact is more resistive than a normal substrate connection in a bulk FET, due to the longer and narrower geometries associated with biasing the body. These electrical characteristics must be factored in when designing with body connections.

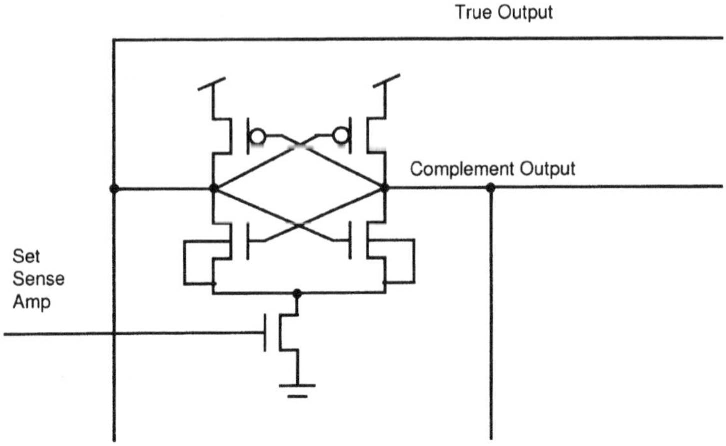

Figure 4-9. Sense amplifier with key body connections.

4. SUMMARY

Silicon-on-insulator technology creates many interesting challenges in memory design and test. It is key to understand and factor in the parasitic bipolar and history effects inherent to memories in SOI. There are greater challenges due to increased loading along a bit line as a function of the data type stored in the cells along a column. The impact of history on cell stability must be considered as well. With careful design and test, performance advantages can be gained and robust circuitry can be implemented in silicon-on-insulator technology.

Chapter 5

Content Addressable Memories

Design & Test Considerations

"To what shall I compare thee?"– Lamentations 2:13

Content addressable memories, or CAMs, determine all or part of their addressing based on data already stored, rather than on an address location. This is the equivalent to recalling a bit of information based on association with another bit of information. It is like asking what color the house is in which a relative lives. The corollary from a random access point of view would be asking the color of a house at a specific street address. It is easier to remember a relative's name than to remember their address location. Just as it is handy and provides significant advantage in human interaction to select data based on a piece of data, it is also helpful to select data stored in a computer memory based on a piece of data already stored in the computer memory. While making this association in a person's mind is easy, the complexity associated with a content-based address in semiconductor memory is incredible.

Content addressable memories are utilized in numerous applications across the industry. One example is a translation look aside (TLB) in a microprocessor [51]. A TLB aids in the mapping of virtual addresses to real addresses. Rather than utilize a full address in a microprocessor, only a piece is utilized. The TLB, however, keeps track of the actual real address. Thus, a TLB is a content addressable memory where part of the data, the virtual address, is used to look up the complete data, the real address [52]. CAMs are utilized many other places in microprocessors. Many of these CAMs are small with only a few entries and yet the complexity of these memories is extremely high.

1. CAM TOPOLOGY

A CAM contains not one but two sections of memory. The first section compares and the second is for reference data storage [53]. These two sections can be referred to as the compare array and data array, respectively, which are shown in Figure 5-1. The compare array selects which section of the data array to read or write. The compare array contains valid bits, the data being compared, and possibly other bits as well. The valid bit, in an entry, determines whether any real data has been entered which should be compared. The other bits can identify if a given data entry has been changed and which entry in the CAM can be erased. When a CAM is full and yet new data needs to be written, one or more entries need to be erased. It is key to know which entry should be the first to be cleared. When a match occurs, in the compare array, with data that is being applied at the inputs of the CAM, a "hit" occurs. The hit forces a word line in the data array to become active. Accessing the data array is identical to SRAM operation, as has already been covered, and will not be discussed further. In some applications the term CAM refers to only the compare array.

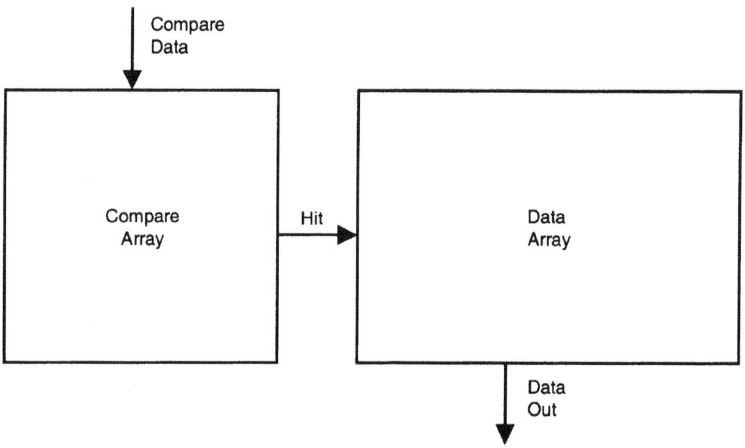

Figure 5-1. CAM compare and data arrays.

One or more entries can have a hit in a given cycle. Depending on the CAM architecture, a multiple hit situation may be tolerable and arbitration logic may be included to select which entry to send to the CAM's output. A priority encoder is often used in this application. If multiple hits indicate an error condition, it should be identified as such at the CAM's output.

Being able to perform a content based addressing function requires a compare to be possible at all of the address locations in a memory. A compare function is logically described in Figure 5-2. For each bit in the compare an XOR operation is performed and fed into a wide NOR gate. If any bit mismatches then its XOR output will go to a "1" and the NOR output will go to a "0", indicating that there was a mismatch. This same compare function needs to be performed on each entry in the CAM's memory.

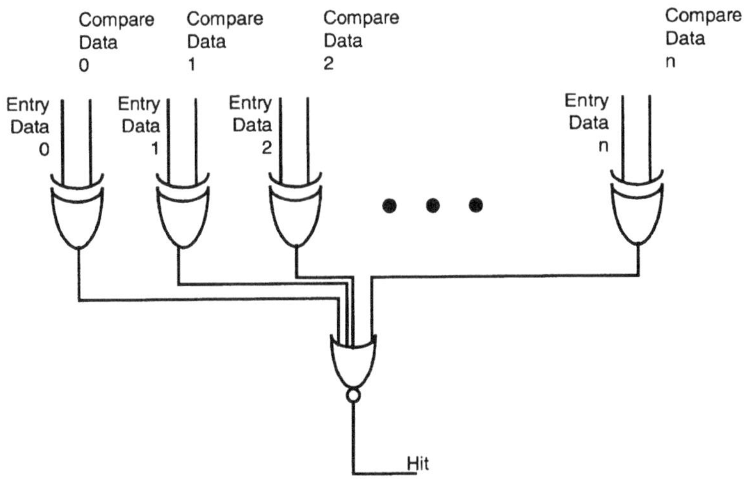

Figure 5-2. Compare function for each CAM entry.

Including an XOR gate for each bit and a wide NOR gate for each entry in a CAM compare array is a prohibitive design choice. Instead, the XOR function is built into the array on a cell basis. Dynamic logic techniques are employed to radically reduce the circuitry required to provide the NOR function. Figure 5-3 shows a cell with a built in XOR function [54]. There are ten transistors, which make up the cell. Six of these are the typical cross-coupled latch and transfer FETs of an SRAM cell. The other four provide the XOR function. If there is a mismatch then either the true side or the complement side causes the hit signal to be discharged. Only one of the stacked NFET pairs can be on at a time, due to data types. If a match occurs on a cell, then neither the true nor the complement NFET stacks discharges the hit signal. All of the compare cells for an entry are attached to the hit signal. If any bit does have a mismatch then the hit signal will be pulled low. Prior to each compare operation the hit signal must be pre-charged high. The NOR function is included in the form of a dot-OR arrangement. This type of a NOR gate is especially effective where a wide set of inputs is required. The CAM can have any number of inputs and still use this type of NOR circuit topology. Typical CAM compare widths are eight to 72 bits.

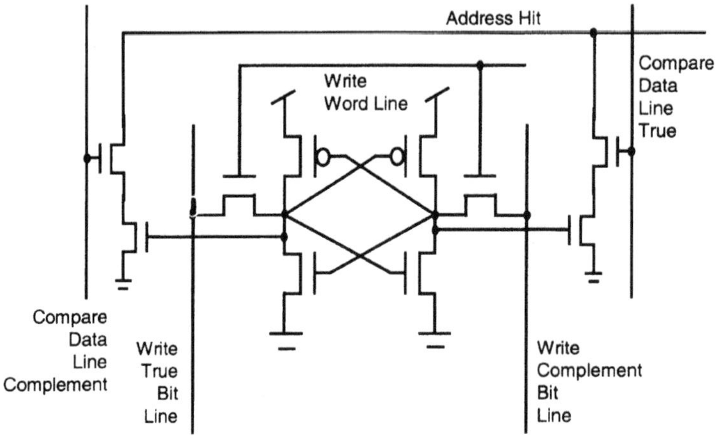

Figure 5-3. CAM cell schematic.

An alternative cell schematic is shown in Figure 5-4, with one fewer transistor employed. If there is a mismatch, the gate of NFET T1 is driven high from one of the compare lines through NFET T2 or NFET T3. This configuration, while being slightly smaller, does not discharge the match line quite as quickly since the gate of NFET T1 is not driven as high in potential.

At the beginning of each cycle the CAM entry address hit signal is pre-charged high. Any mismatch causes the hit signal to be pulled low [55]. The most difficult mismatch to detect is a single bit mismatch. The hit signal spreads horizontally across the compare array and is heavily capacitive. A single bit mismatch requires the hit line to be pulled low through a single NFET stack. Therefore careful timing analysis and simulation must be performed to ensure that worst case conditions still allow a single bit mismatch to discharge the hit signal with sufficient margin. A timing chain exists in the CAM with this topology, which clocks the hit signal into the word line of the data array. This function can be accomplished simply with an AND gate. The clock is ANDed with the hit signal to drive the appropriate word line high in order to read the data array.

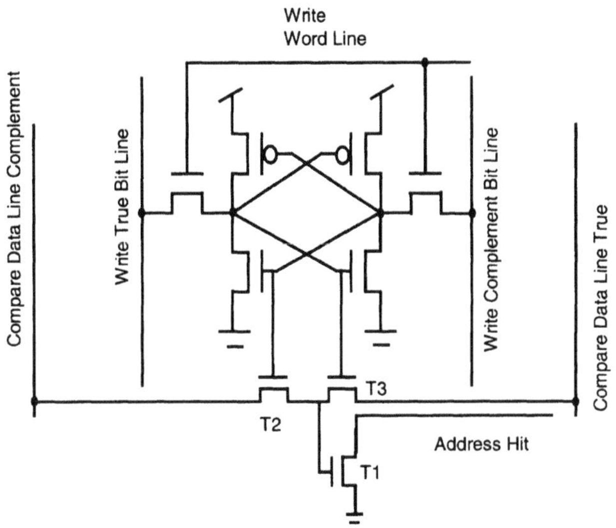

Figure 5-4. Alternative CAM cell schematic.

2. MASKING

In certain applications there are bits which are "don't cares" in either the compare data that is stored in the array or that is applied for compare at the CAM's inputs. These compare inputs can be masked with a masking bit on a per bit basis. In this case, each time a mask bit is set the number of bits that must match is reduced by one. If all of the mask bits are set then all of the entries will indicate a match, which is obviously a useless condition but does illustrate the circuit operation. A cell that can be used with a mask bit input is shown in Figure 5-5. In this case an NFET is added to the stack of those shown previously in Figure 5-3. If a masking bit is set, the inputs to the corresponding NFETs are both driven low. This prevents that cell from discharging the hit line and thus indicating a mismatch. An easier arrangement is handled at the compare bit input circuitry. If a mask bit is set then the compare data lines are both driven low, preventing any of the cells along the column from driving the hit line low. This allows fewer devices in the cell and fewer lines that need to run vertically through the compare array.

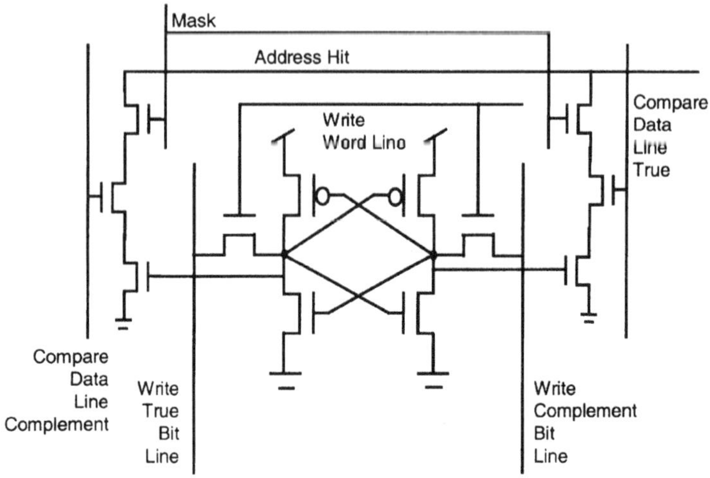

Figure 5-5. CAM cell with a masking input.

A ternary CAM, or TCAM, allows individual bits in entries to be treated as "don't care" bits. A ternary CAM has three states per bit: they are "0", "1", and X. If an X is encountered then either a "1" or a "0" will be considered to be a match on that bit. Since an SRAM cell, the core of a CAM, stores only two states, i.e. "1" and "0", more than one cell is required to store the three states. Figure 5-6 shows a ternary CAM implemented with static cells. Two cells normally have four valid possible states but in the case of a ternary CAM, only three of the states are considered valid.

For a ternary CAM, the two bit cells are referred to as cell X and cell Y. The valid encoded states of these two cells are listed in Table 5-1. Alternatively, some static ternary CAMs use two bits where one bit cell stores the data and the other bit cell stores the mask state. It is arguable as to which technique produces the fastest TCAM.

Table 5-1. Valid states for ternary CAM cell.

X	Y	Data	Mask
0	1	0	0
1	0	1	0
0	0	X	1

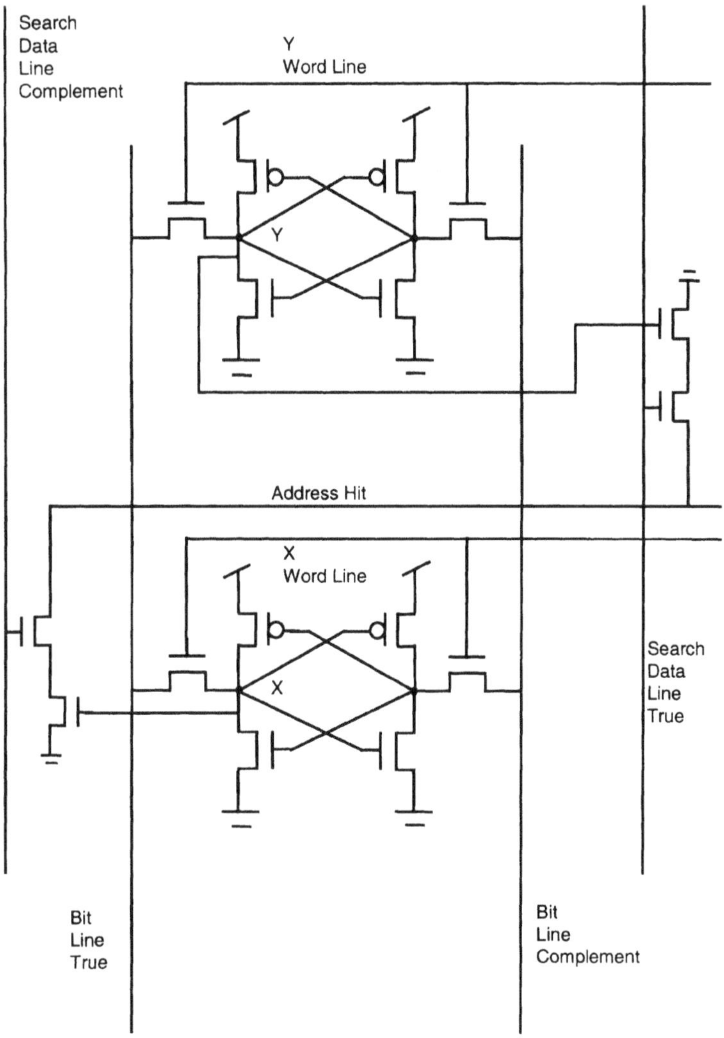

Figure 5-6. Static ternary CAM cell.

A ternary content addressable memory can also be implemented using dynamic bit cells in place of the static ones shown in Figure 5-5 [56]. This memory embodies all of the complexity of a ternary CAM along with the refresh challenges of a DRAM. Figure 5-7 shows a schematic of the dynamic ternary CAM cell. The valid states are the same as shown earlier for the static ternary CAM cell. More will be covered in the next chapter on dynamic memories and the associated design/test issues.

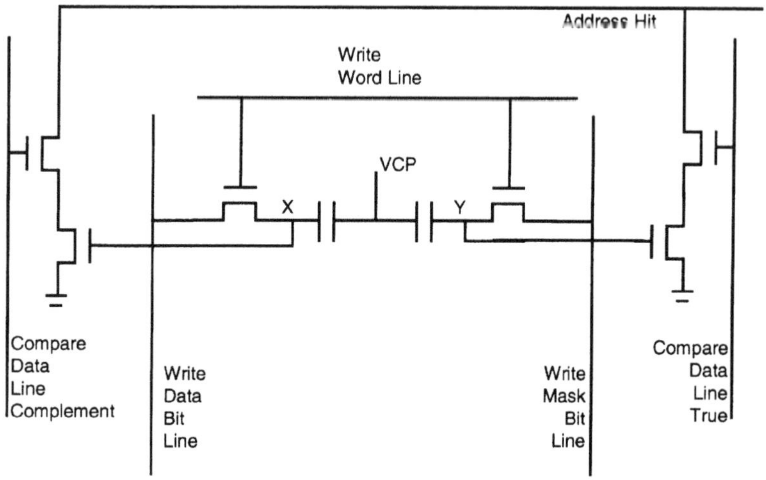

Figure 5-7. Dynamic ternary CAM cell.

3. CAM FEATURES

Occasionally, it is desired to clear a CAM to make room for a complete new set of entries. To accomplish this, the compare and data arrays do not need to be cleared of their information. All that is required is for the valid bits to be set to the clear state. Thus, a flush operation is performed that sets all of the valid bits to zeros. The valid bits are slightly different in that they include a connection line, which allows simultaneous writing of a whole column to zero.

A content addressable memory can utilize compare data to perform some of the addressing and standard random addressing to perform the remainder [57]. In this case the row selection is accomplished through normal data compare, which drives the appropriate word line in the data array to a high state. That word line accesses multiple entries in the data array. Normal addressing is utilized to perform the last stage of decoding, which effectively becomes a column address.

Key point: Content addressable memories include significant complexity in their designs to perform their compare and other functions.

Compare data is typically written into the compare array and only used for subsequent compares functionally. There are good reasons for including a read port on this array, however. During debug and diagnostics it can be useful, at a system level, to read the contents of each CAM array. Thus, it is beneficial to include standard random addressing capability on larger CAMs to accomplish this. Also, during test, being able to read and write the CAM arrays facilitates identification of any failing cells. Without full addressing subtle defects can escape detection and create intermittent field failures.

4. SUMMARY

A content addressable memory is a very useful and yet very complex array. The topology includes an embedded compare function that is spread across the bit cells in the compare array of the CAM. This feature is used to select the remainder of the data to be read or written through the addressing based on matching to this compare. Various masking schemes are possible including a ternary content addressable memory, which allows individual bit cells to either participate in a match as a "1" / "0" or be masked and not participate. CAMs, through their usefulness, will continue to grow in size and will proliferate through other semiconductor chips in the future.

Chapter 6

Dynamic Random Access Memories

Design & Test Considerations

"You've neither sense nor memory." – Long John Silver in *Treasure Island*

Dynamic random access memories or DRAMs are in many ways like their similarly named static random access memory (SRAM) cousins. Decoders select the appropriate x and y location. A word line is activated; a bit line is changed in potential. A sense amplifier takes that small change and develops it into a full rail swing.

In these ways a DRAM is much like an SRAM. In numerous ways, though, a DRAM is radically different from an SRAM. Most notably DRAM cells store their information on a capacitor, which is inherently leaky. A non-volatile memory retains data until it is re-written. An SRAM retains its data until power is removed. A DRAM retains its data for a fraction of a second. Due to the non-ideal nature of the DRAM capacitor, it leaks and after a period of time, loses its data. Thus DRAM memories must be constantly refreshed. Each of the memories discussed in the previous chapters had non-destructive reads. In a DRAM, the act of reading removes charge from a cell capacitor and thereby destroys the data stored in the cell. Testing of a DRAM involves more analog nuances than any other memory type. While SRAMs actively restore their data internal to the cell, only DRAM cell capacitance retains the data stored there. This capacitance is larger than that of an SRAM cell making DRAMs less sensitive to soft errors. Thus, DRAMs are similar and yet unique. The unique aspects will briefly be covered here. For the similar circuit portions, it is recommended that the readers familiarize themselves with the corresponding circuitry described in chapter two.

1. DRAM TRENDS

The density of DRAMs continues to accelerate while the feature sizes continue to shrink. Figure 6-1 contains the predicted reduction in half-pitch size [58]. This value approximates the minimum feature size. These reductions challenge the advanced processing techniques.

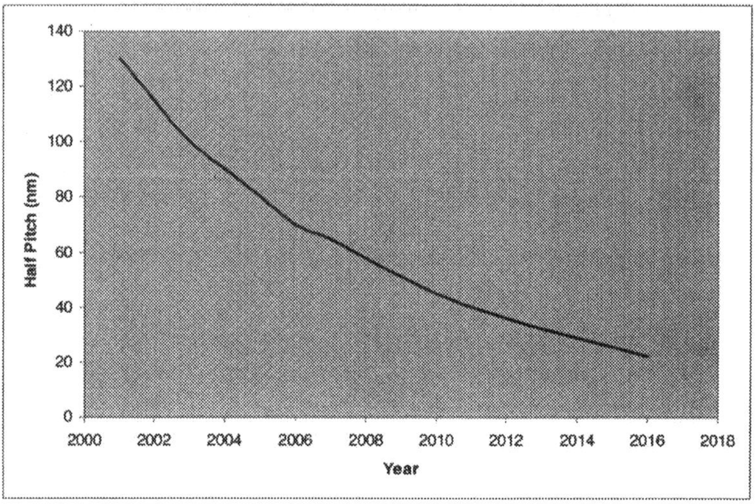

Figure 6-1. DRAM half pitch trend.

The shrinking geometries allow greater memory to be placed on each chip, as shown in Figure 6-2. This increase is needed as customers have an insatiable demand for greater memory storage. The trend will challenge designers to use ever more creative topologies to enable faster access to this vast amount of data. As these greater densities arrive, so to will new fault models. Thus test engineers must be ever vigilant to provide the best test techniques and self-test algorithms to aid in ensuring that customers receive only good product. Furthermore greater amounts of redundancy must be included and the optimal implementation must be assured.

The challenges associated with the design and test of these embedded DRAMs is immense, to be sure. Thus this text primarily focuses on the embedded memories. There are further challenges associated with stand-alone interface design and test for these memories [59]. This topic alone warrants its own text. Stand-alone DRAMs can be asynchronous and operate on an address transition detection, rather than a clock. The row address and column address are often multiplexed through one bus onto the chip [60]. Each of these, while very interesting, will not be covered in this text and the reader is referred to the listed references along with numerous articles in the IEEE Journal of Solid-State Circuits and International Solid State Circuits Conference Proceedings in general.

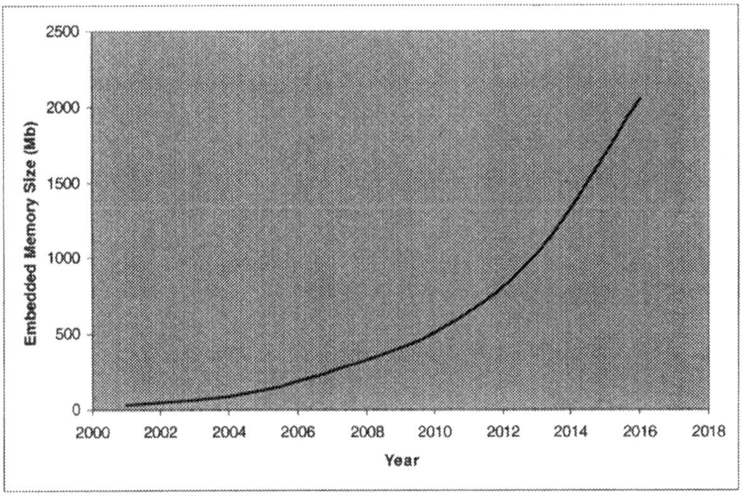

Figure 6-2. Embedded DRAM size trend per chip for mass production.

2. THE DRAM CELL

The DRAM cell has one transistor and one capacitor, as shown in Figure 6-3. The single transistor is activated when the word line goes high. During a read, charge stored in the cell is transferred onto the bit line. Depending on the data stored in the cell, the bit line can be pulled up or pulled down [61]. The sense amplifier evaluates this small change and generates a full rail value of the corresponding data type.

While the schematic shown in Figure 6-3 is normally considered the DRAM cell, there are alternatives. Figure 6-4 shows alternative DRAM cells constructed from multiple transistors. Figure 6-4(a) shows an active pull down device, T1. This structure allows the bit line to be discharged more rapidly. It also causes the read operation to be non-destructive, since charge is not transferred out of the cell and onto the bit line. During a normal one-transistor DRAM read, the cell is "refreshed" by the write-back operation. Since the three-transistor DRAM cell does not require a write-back operation, it still needs a refresh to be performed within the specified time. Obviously using three transistors, two bit lines, and two word lines causes this DRAM structure to be radically bigger than a one-transistor DRAM cell.

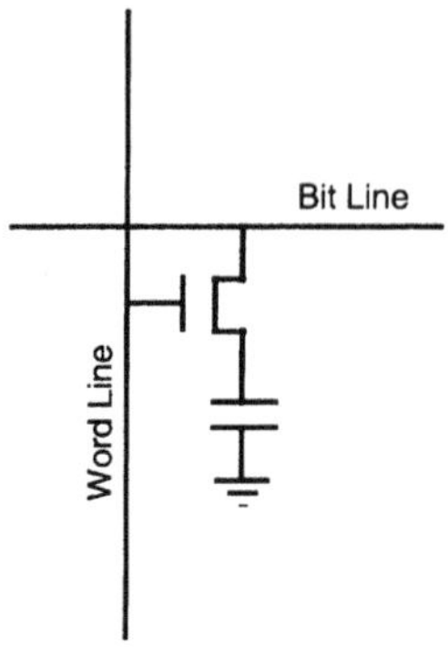

Figure 6-3. One-transistor DRAM cell.

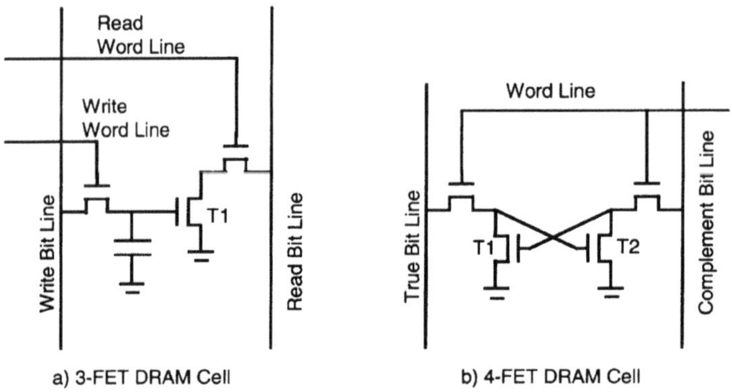

Figure 6-4. Alternative DRAM cells.

Another alternative is to have a four-device DRAM cell, as shown in Figure 6-4(b). This looks just like a four-transistor SRAM cell with the poly load devices removed. One interesting feature of this cell is the means by which refresh is accomplished. The cell does not need to be read and then written back. By simply holding the bit lines high and raising the word line, the desired potentials are restored into the cell due to the feedback arrangement included in the connections between T1 and T2.

3. THE DRAM CAPACITOR

For all but the last DRAM cell discussed, a capacitor is included for storing charge. The capacitor can be formed from a number of structures such as the trench shown in Figure 6-5. In this case a large amount of charge is stored vertically below the active device level on the chip. Significant processing and associated masks are required to implement a trench capacitor and transfer device. The challenging process steps include digging the high aspect ratio trench, growing the vertical oxide and then filling the trench with conducting polysilicon. Trench capacitors do benefit by being processed early, prior to any device, thereby reducing the perturbation to the logic process.

A stacked capacitor may also be utilized as shown in Figure 6-6. In this case the charge is stored above the active device layer on the chip and can benefit from lower series resistance depending on the structure utilized. The capacitor contact does add lithography features and adversely impacts the logic backend wiring. A stacked capacitor also typically only offers half of the capacitance of a trench capacitor.

A third possibility is to utilize the polysilicon, oxide, and diffusion structures that are normally available in any CMOS process to store charge. This capacitor is in the same layer as the active devices for the chip. When this type capacitor is utilized, the amount of charge that can be stored is quite small. Therefore the bit lines need to be short so that a detectable level of signal can be obtained when the cell's charge is transferred onto the bit line. A number of small variations on standard CMOS logic processing have been considered to enhance the capability of reliably storing charge.

For each of these configurations, not only is the capacitance critical but the resistance in series is critical as well. To maintain charge in the cell, the access transistor is designed to be a low leakage device, resulting in a slower device overall. Additionally, the physical structure attaching the FET to the capacitor can add significantly to the total resistance. Since an RC network is formed, the resistance must be understood and appropriately modeled to ensure a good design with sufficient signal development and charge transfer.

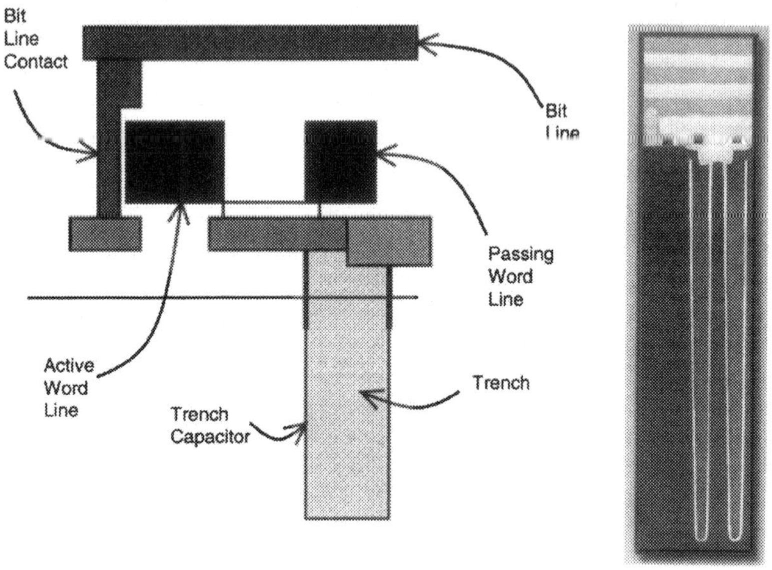

Figure 6-5. DRAM trench capacitor a) diagram and b) photograph (photo courtesy of IBM Corporation / Tom Way Photographer).

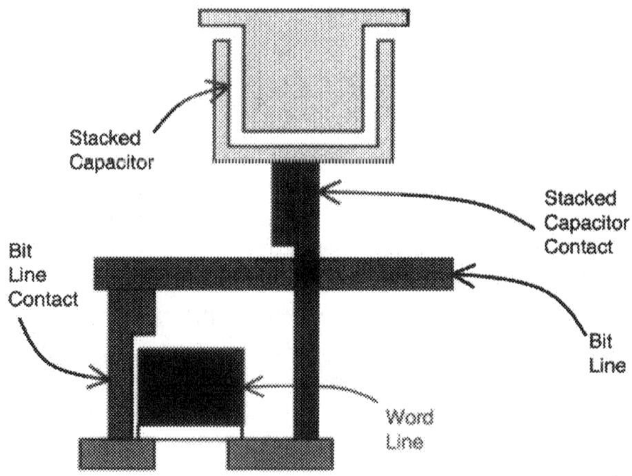

Figure 6-6. DRAM stacked capacitor.

Since a DRAM cell stores charge it is possible for each cell to store more than one bit. If a Vdd or a ground potential is stored in a cell as a "1" and a "0" respectively, it is possible to have intermediate potential values stored as well. In this case 00, 01, 10, and 11 combinations could be stored at zero, one third Vdd, two thirds Vdd, and Vdd. While interesting, cell leakage and data retention issues become more severe for multi-bit storage [62].

4. DRAM CELL LAYOUT

The layout of a DRAM cell is highly optimised to reduce the amount of space it requires. Even the tiniest improvement in individual cell area has a very significant impact. A smaller cell area results in either a smaller overall memory area or, more likely, a larger number of cells in the memory.

Figure 6-7 shows a representation of the bit lines, word lines, and storage capacitor contacts for a DRAM cell. This view is much simplified since the word lines actually bend around the bit-line contacts and the storage-cell contacts to reduce the area further [63]. The word lines run vertically, as opposed to the convention in SRAMs where they run horizontally and the bit lines run vertically.

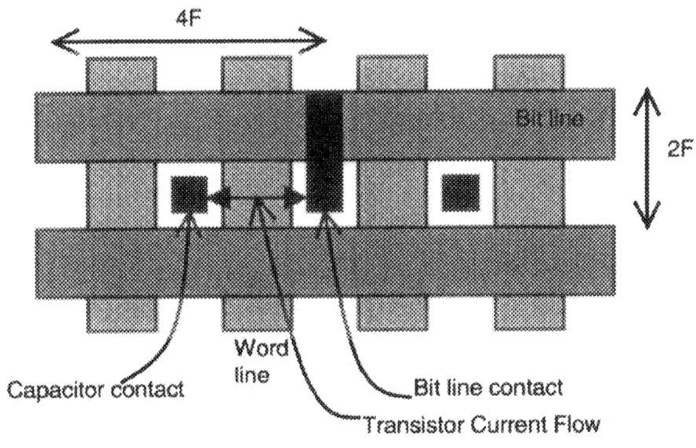

Figure 6-7. Simplified DRAM cell layout.

A single bit line contact services two storage node contacts. A separate word line activates one access device or the other. This cell is referred to as the $8F^2$ cell, which is the industry standard. The derivation of the $8F^2$ comes from the dimensions of the cell. The term "F" stands for feature size. The pitch in the bit dimension includes an F for the width of the bit line and another F for the spacing between bit lines, resulting in 2 F. The pitch in the word dimension includes the width of the word line, half of the width of the bit line contact (since it is shared with the neighboring cell), the width of the storage capacitor contact, the width of the word line that bypasses this cell in order to contact the next one, and half of the space between word-line poly shapes. The total is thus four features or 4F. The overall area is $8F^2$ from the product 2F * 4F.

There are a number of ways to reduce the DRAM cell area to $6F^2$ which largely depend on the overall DRAM architecture. Some of these possibilities will be covered shortly. It is considered that $4F^2$ is the ideal limit for the area of the DRAM cell. To achieve a $4F^2$ area, there is a cross point with only one bit line and bit line space in the y dimension and one word line and word line space in the x dimension generating a small cross-point cell location.

5. DRAM OPERATION

As stated earlier, the read operation of a DRAM involves transferring charge from the storage capacitor of the cell onto a bit line. A multitude of sensing, pre-charge, and bit line configurations have been designed over the years [64]. The discussion that follows describes the operation of a DRAM in general but the specific details of any DRAM design can vary, thereby driving unique design and test challenges.

Figure 6-8 shows the basic internal operating waveforms of a DRAM. Once the specified address has been determined, the appropriate word line becomes active. The bit line which has been selected is decoded and the isolation devices attach a pair of bit lines to a sense amplifier. As seen in Figure 6-8, the word line voltage is boosted to a value above Vdd. The potential needs to be above Vdd in order to adequately perform a write back of a "1" value into the cell through the NFET access device. In the latest technologies, with their associated lower voltages, the boosted word line is also critical to being able to read a "1" from the cell.

Both the bit line and its complement are pre-charged to Vdd/2 prior to the word line becoming active. As the word line is driven high, the access device turns on raising or lowering the bit-line potential as charge from the cell is transferred. At this point the sense amplifier is activated. Only one bit line shifts in potential. The other bit line remains at its pre-charge value, since it accessed no cell, and is used as a reference. When a different word line is activated this bit line will be the one which shifts in potential and the other bit line will be the reference. The bit lines are referred to as bit-line true and bit-line complement but the nomenclature has a different meaning from that of SRAMs. The bit lines service alternating DRAM cells rather than the true and complement halves of an SRAM cell.

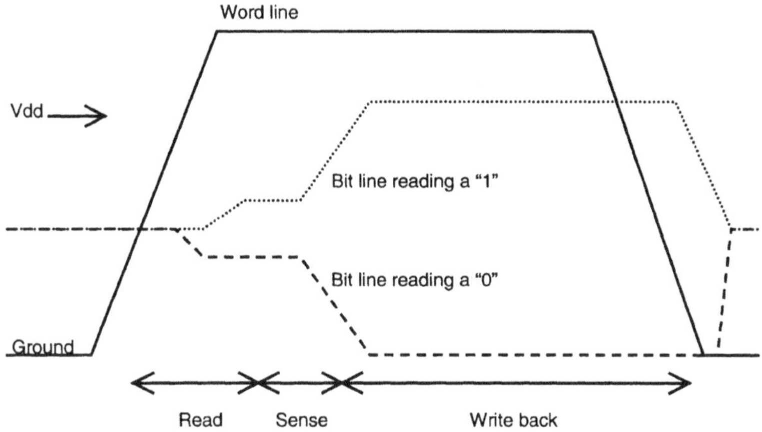

Figure 6-8. DRAM operating waveforms.

Figure 6-9 shows a typical DRAM sense amplifier. It should be noted that the bit-line potentials are applied to the inputs of both the NFETs and the PFETs of the sense amplifier latch. In addition, both pull-up and pull-down devices force the sense amplifier latch to activate. The inputs to these devices are opposite in phase from one another.

Once the sense amplifier is set, the data is re-written to the cell. For every read there must be a write-back operation. Thus Figure 6-8 shows the read, sense, and write-back operations being performed for a typical DRAM topology.

Key point: DRAM operation involves more analog nuances than any other type of memory.

The bit-line arrangement is critical to the overall area of the DRAM and to the amount of noise that can be tolerated. A folded bit-line topology is shown in Figure 6-10 where the bit-line pair accesses an adjacent pair of DRAM rows. The two bit lines are adjacent to one another, allowing bit line twisting that enables most noise to be common mode. Thus the folded bit line approach is less noise sensitive which is the reason that it has very broad usage. The sense amplifiers shown here are shared by both the left hand bit lines and those on the right. The word line drives are shown to be interleaved. A folded bit line requires that each line only contact every other cell, resulting in a larger physical area.

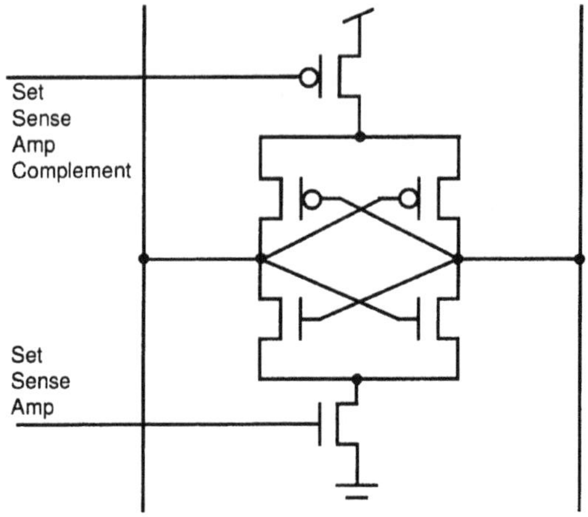

Figure 6-9. Typical DRAM sense amplifier.

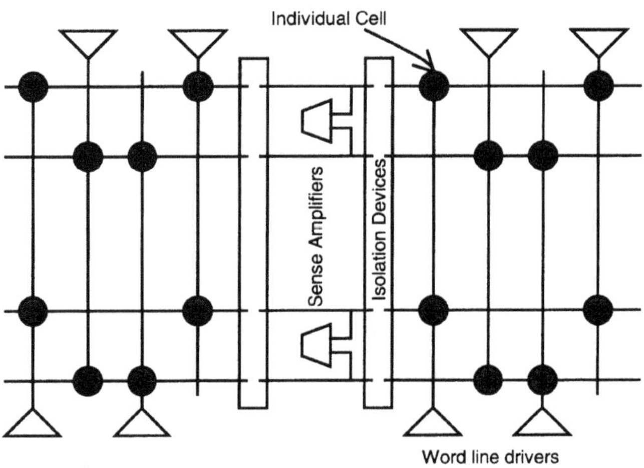

Figure 6-10. Folded bit line arrangement.

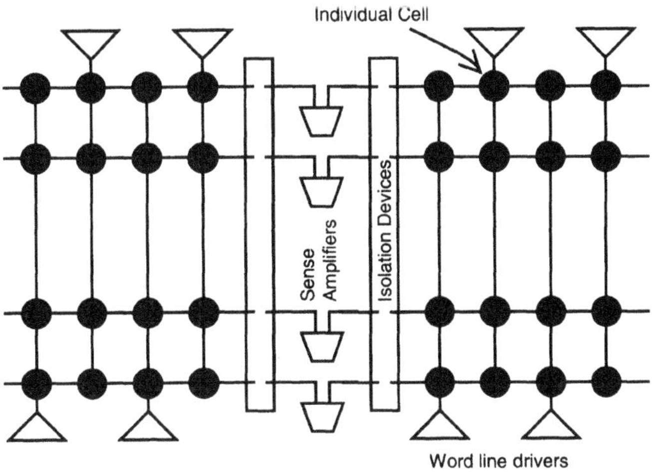

Figure 6-11. Open bit line topology.

An open bit-line configuration allows the line to contact every cell that it passes over. The x dimension of the cell can be reduced to 3F with the resulting overall area being $6F^2$. Any noise is now differential, which is often not tolerable. Other methods for keeping the area of the DRAM down include rotating the array, including a vertical transistor, and tilting the active area [65,66].

6. CONCLUSIONS

Dynamic random access memories are marvelously dense and complex circuits. They are ideal memories for a certain class of applications. It is important to understand the differences that make DRAMs unique from other memories so that they are used in the correct applications. These unique aspects also require the proper fault modeling and test development. Since the analog effects are more severe in DRAMs, more test patterns and special voltage potentials need be applied and injected to thoroughly test the memory. Certainly data retention must be well tested since all cells naturally bleed charge from the DRAM capacitor and the data retention duration needs to meet specification.

Chapter 7

Non-Volatile Memories
Design & Test Considerations

"I'll note you in my book of memory." – from Shakespeare's *Henry VI*

Non-volatile memories range from simple older technology to memories in heavy, industry-wide usage today through the most promising possibilities for future memories. Non-volatile memory is key to the industry because some information must be retained even when power is removed from a chip. Non-volatile memories enable this operation.

1. ROM

A read-only memory contains a fixed set of data for the life of the chip. Its programming is part of the design. Typically micro-code or some other permanently fixed set of instructions or data is programmed in the ROM. The logic design includes an array identifying which bits are to be zeros and which are to be ones.

These ones and zeros are implemented in a simple cell structure like that shown in Figure 7-1. There are four cells illustrated along the column. Bit zero, one, and three are defined to be in the "0" state while bit two is in the "1" state. The lack of an FET in bit position two prevents the bit line from being pulled low by a small amount when word line two is driven high [67]. Since the bit line remains high, a "1" is detected by the sense amplifier [68]. There are a number of ways that this type of ROM programming can be physically implemented. A transistor can be missing or just the contact to the bit line for that transistor can be removed. A difference in diffusions can be implanted to turn off a "1" state cell while leaving on a transistor that corresponds to a cell's "0" state [69]. Alternatively, a diffusion can be implanted allowing a transistor to be turned off, corresponding to a "1" in a stack of transistors. If this implant doesn't exist then the transistor is bypassed, providing a "0" state in the cell. Regardless of the method, a "1" or a "0" exists in a ROM cell based on a difference in the design.

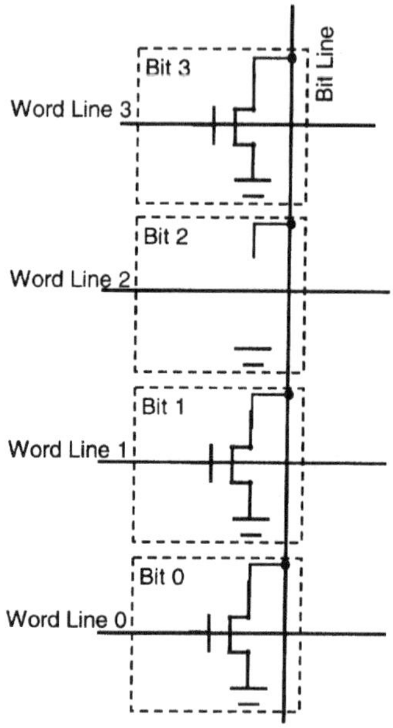

Figure 7-1. Four ROM bit cells along a column.

The ROM is the most non-volatile of the non-volatile memories since the state of the cells can never change. Therefore the ROM is the simplest of structures to design and relatively simple to test.

2. EEPROM & FLASH

After early ROM memories, there were programmable read only memories (PROM) and then erasable programmable read only memories (EPROM). A PROM was one time programmable. An EPROM could be programmed multiple times but could only be erased with the aid of ultra-violet light being applied to the chip. The next logical step was an electrically erasable programmable read only memory. This type of memory goes by the acronyms of EEPROM or E^2PROM. Flash memory, is an enhancement of EEPROM which allows erasing a large memory block in much less time. Many of the design and test considerations are similar between EEPROM and flash allowing them to be discussed here together.

EEPROM and flash are a major boon to the industry because their non-volatile memory is available to be read infinitely and to be electrically re-written numerous times. Erase and electrical re-write provides tremendous freedom in packaging since the application of ultra-violet light is no longer needed to erase the memory. EEPROM can be erased a byte at a time [70] while flash can be erased in large blocks, and possibly the whole memory, at a time.

To accomplish these functions some unique memory circuit arrangements were developed. Figure 7-2 shows a typical schematic for an EEPROM cell, with the memory transistor at the bottom and the access transistor electrically between the bit line and the memory transistor. A variant on this allows the memory transistor to be between the access transistor and the bit line. Since an EEPROM can be erased a byte at a time, two transistors are always required in a cell. To read an EEPROM cell, the word line and the select line must both be brought high. If the cell transistor is not programmed, the threshold voltage is not elevated and the bit line discharges. If the cell has been programmed then the bit line remains high.

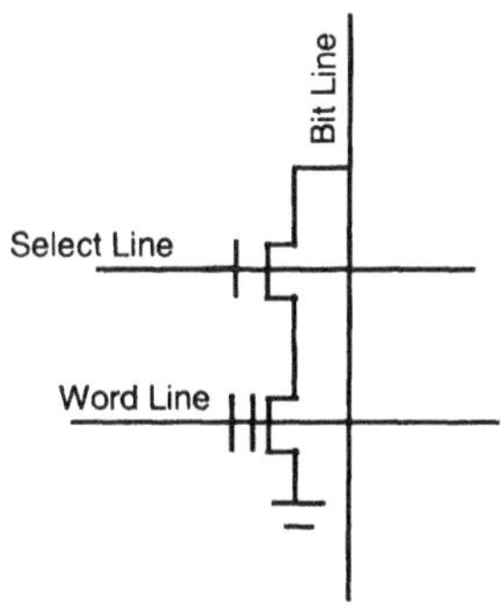

Figure 7-2. EEPROM cell schematic.

A flash memory cell can be either of the NOR or NAND variety, as shown in figure 7-3. The NOR cell can simply have one transistor as shown with the source line going low and the gate going high in potential, to perform a read. A NAND flash memory has multiple programmable cell transistors, typically 16 in a stack, along with two select FETs [71,72]. To read a NAND flash cell, all of the word lines are activated except for the addressed word line, which remains low [73]. The two select FETS, at the top and bottom of the stack, are activated. Depending on the state of the program transistor, the bit line is pulled low. A NAND flash memory is denser than alternatives, especially with the reduced number of required contacts, since diffusion can be adjacent to diffusion without any intervening contact.

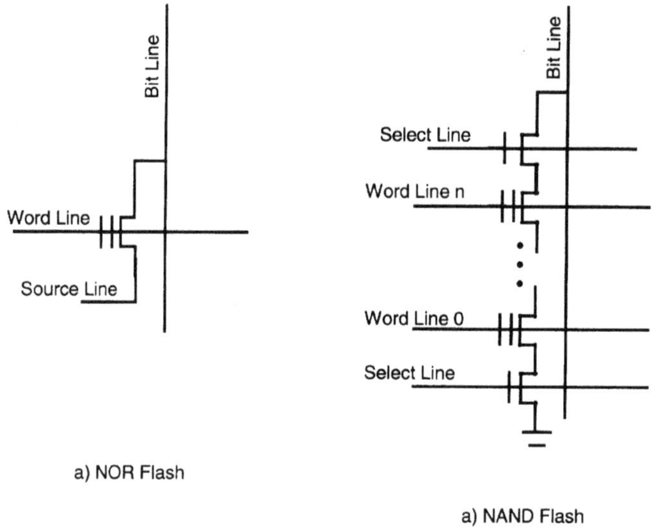

Figure 7-3. Flash cell configurations.

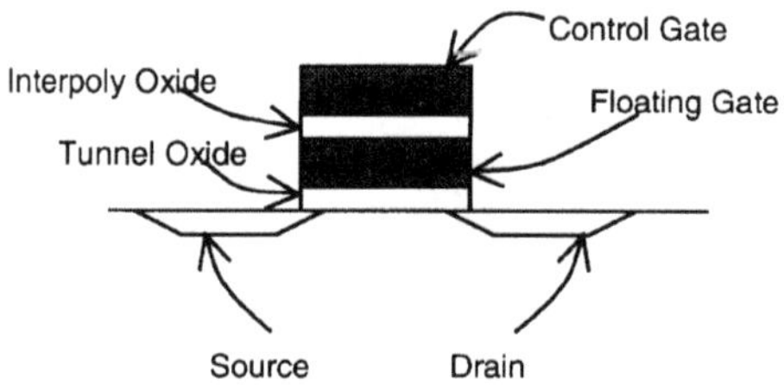

Figure 7-4. Program memory transistor.

A simple program transistor cross-section diagram, which is applicable for both EEPROM and flash memory, is shown in Figure 7-4. The floating gate is isolated from the channel by a very thin layer of oxide and isolated from the control gate by a thicker layer of oxide.

Programming a floating gate involves making its potential negative which has the effect of increasing the threshold voltage of the device. There are different techniques for programming and these will be summarized shortly.

Erasure of a program transistor is accomplished by Fowler-Nordheim tunneling. To cause this type of erasure, a large electric field is applied across the floating gate [74]. Two methods for generating the high field are shown in Figure 7-5. The left erasure method is referred to as source-side erase while the method on the right shows the channel erase technique. The arrows indicate the direction of movement for electrons. Other alternatives for erasing are possible including a source erase by applying a positive 12 volts on the source with the gate being grounded. Fowler-Nordheim tunneling is used for erasing both flash and EEPROM memories.

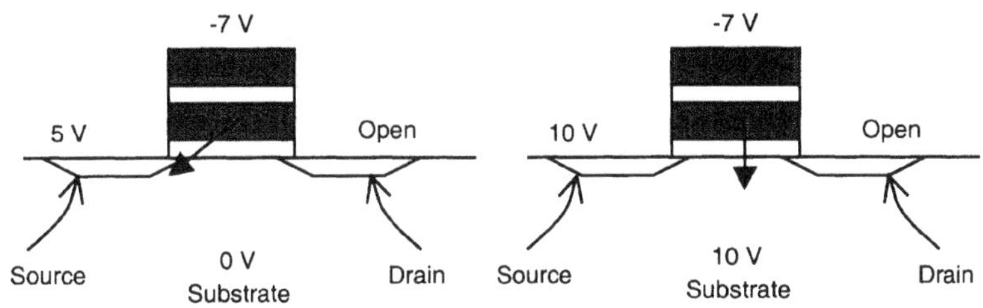

Figure 7-5. Fowler-Nordheim tunneling to erase a memory transistor.

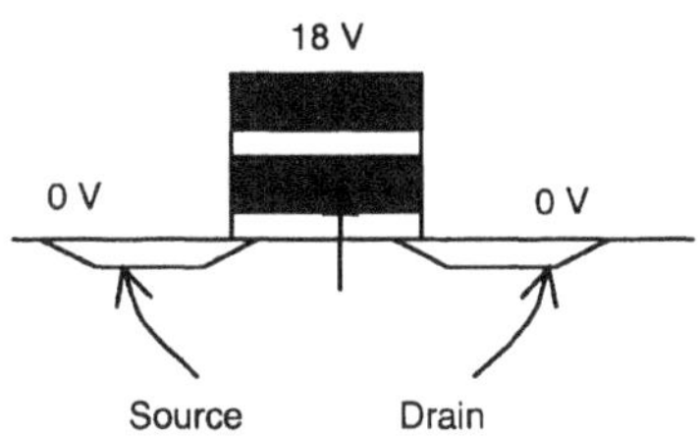

Figure 7-6. Fowler-Nordheim programming of memory cell.

Programming is accomplished by either Fowler-Nordheim tunneling or by hot electron injection. Fowler-Nordheim tunneling to program is performed by driving the gate high in voltage while the source and drain are low, as shown in Figure 7-6. Electrons tunnel through the tunnel oxide driving the floating gate negative. The negative floating gate makes the threshold voltage of the device more positive, thus making the device harder to turn on. Other voltage configurations are possible while still using Fowler-Nordheim tunneling to program a cell.

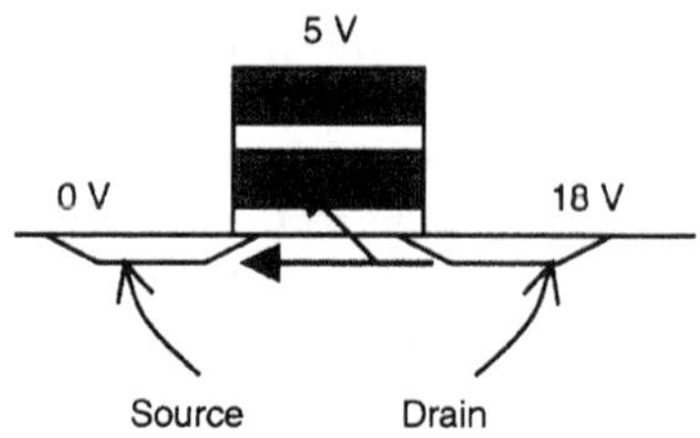

Figure 7-7. Programming EEPROM memory FET.

The second overall method for programming these type memory cells is hot electron injection. A write is performed by injecting hot electrons into the floating gate of the cell, through the application of a high voltage on the drain and a "1" condition on the gate as shown in Figure 7-7. The high voltage on the drain is significantly higher than the normal Vdd and may be as much as 18V or more, depending on the technology and specific memory topology. This programming voltage is referred to as Vpp. The elevated voltage causes a large lateral electric field between the source and drain of the FET, indicated by the horizontal arrow. The transverse electric field between the channel and the floating gate causes some of the electrons to embed themselves in the gate. The electrons travel through the oxide which is referred to as a "tunneling oxide" even though the transfer mechanism is not tunneling but rather hot electron injection. These embedded electrons alter the threshold voltage of the FET, with the threshold voltage indicating whether the cell has been programmed or not. The programmed state can be either a "0" or a "1", depending on the topology of the memory.

Both Fowler-Nordheim tunneling and hot electron injection programming techniques are used in both EEPROM and flash memories, leading to some confusion. For the most part EEPROM utilizes Fowler-Nordheim tunneling for programming. NAND flash also normally uses Fowler-Nordheim tunneling [75]. NOR flash generally utilizes hot electron injection for programming. There are numerous subtle variations in programming and erasing techniques for versions of EEPROM and flash memories. Only the broadest categories have been summarized here.

An alternate cell cross-section, shown in Figure 7-8, is the split gate device. A control gate can make a step, covering the entire floating gate or a control gate and a program gate can be utilized as shown. The addition of a program gate facilitates the use of lower voltages for program and erasure [76]. While programming the cell the program gate is elevated but during erasure both the program and the control gate are driven to a negative potential.

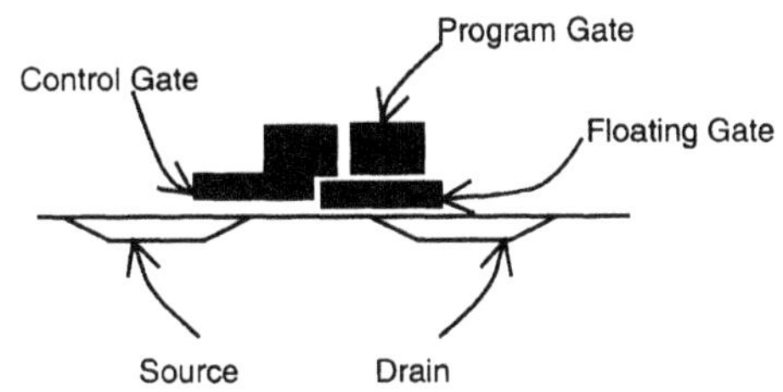

Figure 7-8. Alternative transistor cross-section.

An interesting feature is that these non-volatile cells can store more than a single binary value. Based on the amount of modification to the threshold voltage, multiple bits can be stored in a single cell [77,78].

While these memory cells are very useful there are some significant limitations. There is a limited number of erase and write cycles, normally in the range of 100,000 times. Also, these memories read rapidly but have long erase and program times. Some transistors must exist on the chip that can handle the elevated voltages and steer them to the cells which need to be written or erased [79]. The handling of elevated voltages definitely creates greater complexity.

Flash and EEPROM memories can have program disturbs where the elevated voltage on the gate and drain influence cells other than the one intended to be programmed [80]. Read disturbs are also a concern where all of the cells along a bit line are candidates for disturbing the data being read from the intended cell. If a bit line is intended to remain high but is inadvertently discharged by a defective FET, a read disturb has occurred. Beyond disturb concerns, it is important to test these non-volatile memories to ensure that there is good oxide integrity and that data retention is not a problem.

3. THE FUTURE OF MEMORIES

Entirely new memory types are not developed very frequently but, based on the last few years, a number of non-volatile contenders threaten to radically change the future of memory. ROM, Flash, and EEPROM each have their place but several non-volatile memory types are vying for the future with much broader application than the non-volatiles of today. These new memories have some aspects which reflect the endurance, performance, and power of SRAMs and DRAMs but are non-volatile. It is uncertain as to which technology will dominate since each has strengths and weaknesses. The possible benefits to existing applications and the contemplation of new applications based on the enhanced capabilities of these memories can provide a huge shift in the computing and electronics market as a whole.

3.1 FeRAM

There has been much published on the ferroelectric random access memory, also known as an FeRAM or FRAM. The method by which an FeRAM stores information is through an electrically polarizable material which maintains a certain amount of polarization even without an electric field present. Thus, an electric field is applied to create a polarization which is then read back at some later point in time.

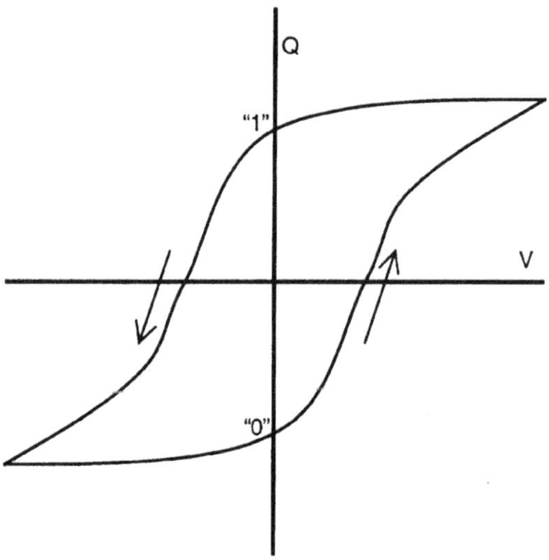

Figure 7-9. Polarization hysteresis curve.

The key material component of an FeRAM is, paradoxically, not ferrous in nature. There are a number of polarizable materials which can be utilized in a FeRAM. The most common of these is PZT which has a molecular description of $Pb(Zr_xTi_{1-x})O_3$ [81]. Other materials that can be used in FeRAMs are referred to by the acronyms SBT, BST, and BLT [82]. The polarization of these materials follows a hysteresis loop [83] that is shown in Figure 7-9. When voltage is applied, the material polarizes, following one of the arrows in the diagram to reach a polarization of a "1" state or of a "0" state. When voltage is removed, a residual or remnant polarization charge is detectable as a value of Qr or –Qr, with the assignment of "1" or "0" to these values arbitrarily.

A ferroelectric capacitor, which retains the residual polarization, functions in many ways the same as a DRAM capacitor, as can be seen from the FeRAM cell schematic in Figure 7-10. This schematic shows a single device and single capacitor [84] giving the cell its 1T1C name. Historical memories with a 2T2C differential cell were much larger. In order to make FeRAM viable for large scale use a smaller cell was required. Most 1T1C FeRAMs have a folded bit line architecture [85] like that described in chapter six for DRAMs. This approach keeps noise down but does have an area penalty.

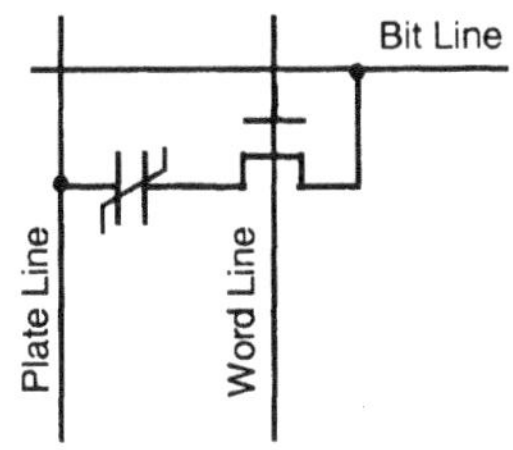

Figure 7-10. FeRAM cell schematic.

Each time that a cell is read, the cell's polarization must be rewritten since a read is inherently destructive for FeRAMs. Because of the destructive read, it is very important that an FeRAM have sufficient endurance to handle all of the reads and writes without detriment to its operation. Endurance is limited by hysteresis fatigue which demonstrates itself with a flattened hysteresis loop and reduced residual polarization. Sense schemes can employ a reference cell which is cycled on every read operation. This reference cell is the most likely to have problems due to its limited endurance. Current commercial FeRAMs have a fatigue limit of 10^{12} operations. An FeRAM can also develop an imprint [86] where the hysteresis loop shifts vertically to give a preference for one data type over another. Polarization charge can also decrease over time causing a slow loss of the cell's data [87]. Bit lines are typically pre-charged to ground [88] and a reference bit line is a relatively small voltage that distinguishes between a "1" and a "0" state being read from the array [89].

3.2 MRAM

In antiquity computer memories utilized magnetic core elements. These cores were large tori (donut shaped objects) which could have their magnetic fields reversed based on whether they were storing a "1" or a "0". One of the leading possibilities for future non-volatile memories is not large tori but is a magnetic material with reversible polarity. Magnetoresistive random access memory (MRAM) contains material which changes in resistance depending on the direction of its magnetic polarization [90,91]. This change in resistance enables it to store binary values. There are a number of materials and memory layout arrangements that have been used in MRAM and the review here is just a brief overview. Other configurations can be examined in the reference material.

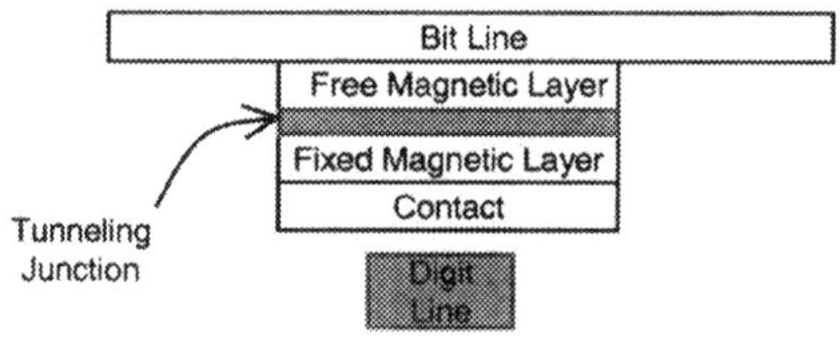

Figure 7-11. MRAM material sandwich.

The basis of MRAM is a thin material with reversible magnetic polarity. Figure 7-11 displays a "sandwich" material that is at the core of each MRAM cell. It contains a contact top and bottom for electrical evaluation of the sandwich's resistance. In between are a fixed magnetic polarity material and a variable magnetic polarity material. The fixed material is often referred to as having its polarity "pinned" in a specific direction [92]. This pinned polarity material aids in enabling a change in the polarity of the variable material. It is also possible to use two variable polarity materials but one is fabricated to be much more resistant to change. When one of the materials is pinned the sandwich is said to be of the "spin-valve" type whereas when both materials can change it is referred to as a "pseudo-spin-valve" type [93].

In between these two magnetic materials is a very thin tunnel junction layer. This sandwich often goes by the name of magnetic tunnel junction or MTJ [94]. Another conductor is placed below the sandwich but is electrically isolated from it. Current is passed through the top line and the conductor below the sandwich to induce a magnetic polarization in one direction or the other. Reversing the current reverses the polarization.

Some have attempted to use this type material without a transistor but with limited success. Most employ an MRAM cell like the one shown in Figure 7-12. The bit line and the digit line form a 90 degree crossing. In between these two lines the sandwich material is placed and allowed to be polarized by currents through these two conductors. During a read, the word line is activated, drawing current through the bit line and sandwich material [95]. Based on the resistance of the material, which is a function of the direction of the magnetic polarization, a "1" or a "0" is detected from the cell. A reference bit line attaches to a MRAM reference cell which has not had its sandwich material polarized [96].

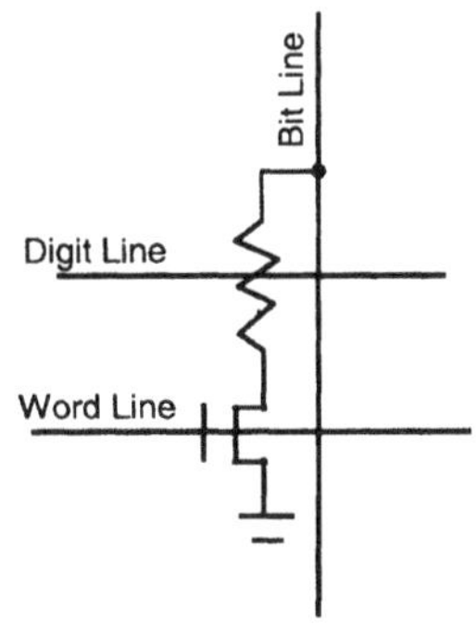

Figure 7-12. MRAM cell schematic.

While MRAMs have the density of DRAMs with significantly lower power, there are other benefits as well. An attractive feature of MRAMs is that they are impervious to soft errors from alpha particles and cosmic rays. Further there is no known limitation on the number of cycles which they can be put through and therefore have no problem with endurance.

3.3 Ovonic

The memory storage attributes from CDROMs and DVDs has been extended to semiconductors in the form of ovonic memories. Ovonic unified memory or OUM utilizes phase changes between amorphous and polycrystalline states of chalcogenic materials to distinguish between a "0" and a "1". By generating a small amount of localized thermal energy, the ovonic memory cell can be set or reset [97].

An ovonic memory cell is reset when it is in its amorphous state, which is higher in resistance. The cell is in its set state when it is in its polycrystalline state, which is lower in resistance. A short electrical pulse puts the cell into its reset state while a lower but slightly longer pulse puts the cell into its set state. The cell can be set and reset at least 10^{12} times and it can be read an infinite number of times.

Figure 7-13 shows a schematic of the ovonic memory cell. In standby, the word lines are at Vdd and bit lines are at ground. This state is the same for unselected rows and unselected columns [98]. No high voltage transistors are needed with ovonic memories. During a write, the selected word line is grounded and the selected bit line is biased to an intermediate voltage.

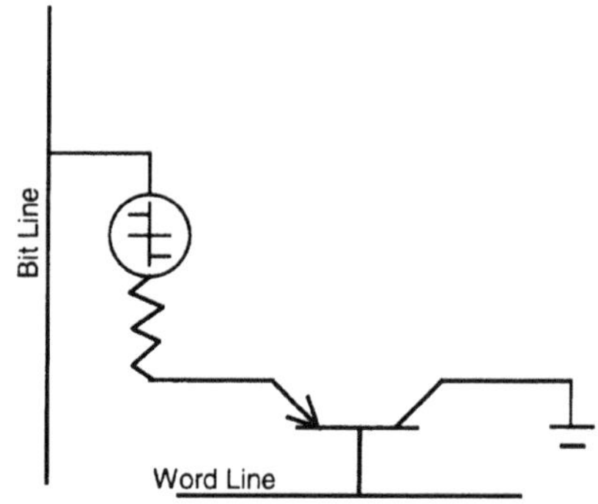

Figure 7-13. Schematic of ovonic memory cell.

Ovonic memories have significant advantages in that they have a non-destructive read, can be read infinitely, and cannot be disturbed by soft error radiation. This technology is also highly scalable to future lithographic limits.

3.4 And Beyond

Future new memory types certainly seem to be non-volatile. With the three new entries in this section, one of them or a slight modification of one will probably be the dominant non-volatile memory in the near future. Beyond this, there will be new memory technologies, some based on silicon and some based on carbon-based systems. One of the latter that seems as strong as any possibility is a carbon nanotubule [99]. This nanotubule can contain a buckyball with a charge on it. Based on the position of the buckyball a "1" or a "0" state can be realized.

4. CONCLUSIONS

Non-volatile memories provide significant advantage in the industry today. Furthermore, they are poised to provide a huge shift in the entire electronics industry as greater and essentially infinite endurance becomes available. The larger endurance will solve one of the primary test problems of how to thoroughly test a non-volatile memory without diminishing its usable life in the field. Nonetheless, the unique design aspects will drive unique requirements in the fault modeling and testing of these future memories.

Chapter 8

Memory Faults

Testing Issues

"This isn't my fault!" – Calvin & Hobbes

1. A TOAST: TO GOOD MEMORIES

The objective of memory design and memory test is to produce good memories and to ship them to the customer, thereby resulting in a profit. The objective of designing a memory is to result in a robust circuit which is not marginal. Then when the memory is tested, the specification is not being assured but rather the circuit is being examined for manufacturing defects. Thus one wants to design a memory with margin so that the circuits can be exercised outside the power supply window by a small, or even a large amount, and have functionality maintained. This functionality may continue at a slow rate if the power supply is reduced to a lower voltage, yet logically it provides the correct binary values. Similarly, if the part remains functional at elevated temperature and voltage significant robustness is provided during field operation. Designing for burn-in conditions can thereby enhance robustness [100].

The purpose of testing, be it the testing of stand-alone memories, through-the-pins testing, or by self testing is to detect manufacturing defects. Characterization, on the other hand, is defined as examining the margins of a design and performing design verification. Characterization looks at marginality with respect to the voltage, temperature, timing, and any other environmental conditions. This is done to verify that the design is, in fact, a good design. When a chip design is first fabricated it should be characterized to ensure that the design functions and that it is robust.

The objective of this text is to discuss testing and examining for manufacturing defects, as opposed to characterization. There will be insights that become obvious on memory characterization issues, however that is not the purpose of this writing.

If the testing of memories is examined, in contrast to the testing of logic, one will notice a markedly different approach. In the testing of logic, there are regular references to fault coverage. This fault coverage is with respect to a given set of fault models. The most common of these is the *stuck-at fault model*. This and the other fault models are discussed in detail later. Writers regularly refer to coverage, be it 90, 95, 99, or 99.9%. Whatever the coverage requirement is defined to be, it is considered to be sufficient for that generation of chips.

In memories, other fault models must be considered. Anyone who approaches memory testing with the view that considers the stuck-at fault model as sufficient, regardless of the coverage percentage, is exceedingly foolish. The stuck-at fault model is not the only way by which memories can fail. Most memories may not fail to the stuck-at fault model and yet numerous faults can still exist and need to be culled out. The objective in testing is to discard *all* defective memories that will fail in the field, not just eliminate those of a certain fault model.

Key point: Many fault models must be covered during memory testing.

While the testing of memories for many fault models is required, the test time cannot be allowed to become prohibitive. Testing of logic requires applying patterns through scan chains with the associated scans between each test. Testing a memory can be accomplished with cycle-after-cycle testing with a new test applied each cycle. Thus, even though many fault models are tested the overall test time is quite short. If there is logic or analog circuitry on chip, the test time will be dominated by testing these and not the memory. It is rare that the test time is driven primarily by memory testing for any system on chip (SOC).

The objective is to generate good memories. A good memory is one that is fault free. Thus, a memory designer or memory test engineer should be able to gladly toast "To good memories," meaning it in the semiconductor sense.

2. FAULT MODELING

Before progressing further into the specifics of test, it is important to discuss the possible range of fault modeling. A fault model is a description of the way something can fail. This description can be done at varying levels of abstraction, as shown in Figure 8-1. Abstraction is key to prevent getting too many details involved in the analysis and test generation [101]. There is a significant risk, however, that abstracting away too much obscures the real details that require understanding to ensure good testing and therefore provide high quality memories.

The highest level of abstraction utilizes behavioral modeling; VHDL or Verilog is often used for this. Functional modeling identifies the circuit as a black box [102]. The inputs and outputs are defined along with their function but visibility to the inner workings of the memory is not provided. Functional fault modeling historically has been the primary level for memory test. Greater detail is gained at the logical level, where logic gates are understood. Below this is the electrical level of fault modeling where the transistor level operation is perceived. For much of memory testing, the electrical level is required or else inadequate tests result. Below this level is the geometrical level of fault modeling. The actual shape placement on the chip is understood at the geometrical level. Certain faults require an understanding of geometric adjacencies to make optimal or even adequate tests, as is the case with multi-port memories. Even in single port memory cells, however, the susceptibility to certain faults is definitely a function of geometric arrangement of the transistor connections. Below the geometric level, the actual paths which electrons and holes follow could be described but the value would be debatable.

Different fault models require different abstraction levels. A single level of abstraction, while desirable from a simplicity point of view, is not attainable. Certain faults require differing levels of detailed understanding, such as the circuit or geometric level.

The thought of coverage percentages is an unacceptable thought when testing memories. For example, on a gigabit chip a “once in a million event” occurs one thousand times [103]. Memories are very dense and therefore have many adjacent structures, all of which need to be tested. Many think of memory testing as following a functional type of approach. Others refer to memory testing as being algorithmic. The opposite is structural testing and it is typically considered in logic test. In Figure 8-2 an XOR, NOR, and OR gate combination is shown. The testing of these gates requires a specific set of structural patterns in order to identify the stuck-at structural faults. A stuck-at test checks to ensure that no gate input or output is stuck-at a "0" or a "1" state.

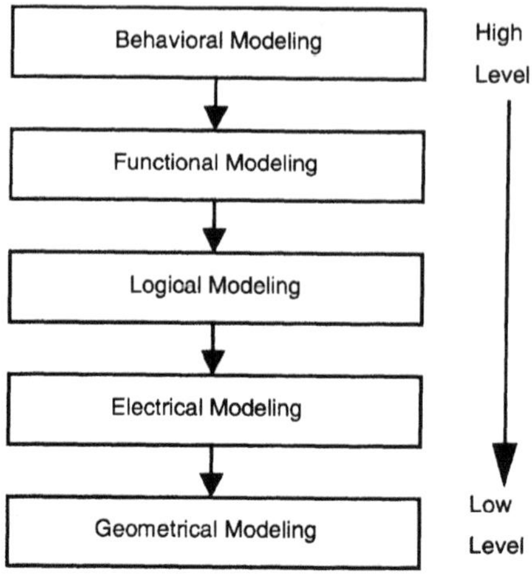

Figure 8-1. Abstraction levels for fault modeling.

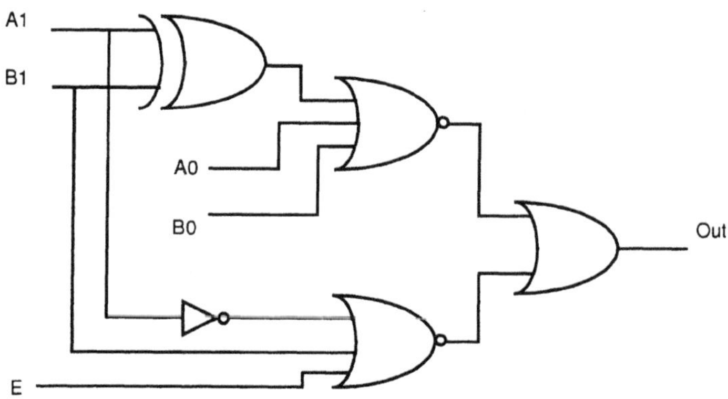

Figure 8-2. Example combinational circuit.

For a stuck-at fault model, each 2-input AND gate requires three test combinations, as does a 2-input OR gate. These are shown in Figure 8-3. For the 2-input XOR gate, all four data type combinations are required.

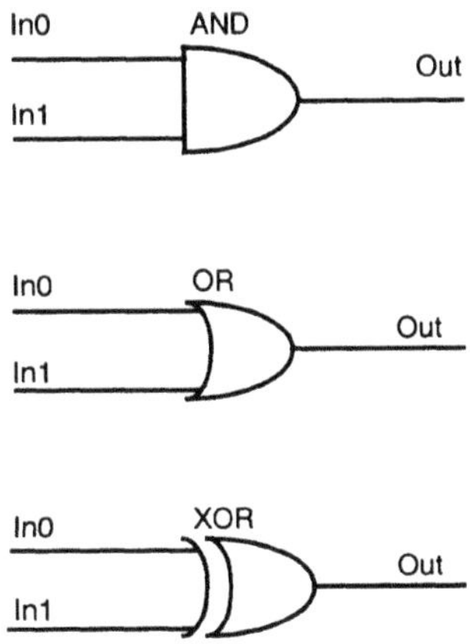

Pattern No.	In1	In0	Out
1	0	1	0
2	1	0	0
3	1	1	1

Pattern No.	In1	In0	Out
1	0	0	0
2	0	1	1
3	1	0	1

Pattern No.	In1	In0	Out
1	0	0	0
2	0	1	1
3	1	0	1
4	1	1	0

a) Simple gates

b) Stuck-at patterns

Figure 8-3. Simple gates (a) and their required stuck-at test patterns (b).

A five input circuit as shown in Figure 8-2 would require 32 input combinations to exercise all possible patterns. However, through ATPG subsumation, IBM's TestBench tool fully tests the circuit for all possible stuck-at faults with only eight input combinations, as shown in Table 8-1 [104].

Table 8-1. Patterns to catch stuck-at defects for combinational circuit.

	A0	B0	A1	B1	E	Out
1	0	0	0	0	0	1
2	1	0	0	0	0	0
3	0	1	0	0	0	0
4	0	0	1	0	0	0
5	0	0	0	1	0	0
6	0	0	1	1	0	1
7	1	0	1	1	0	0
8	0	0	1	0	1	1

By examining the gate level schematic shown in Figure 8-2, it can be seen that the exact transistor level representation has been avoided. This is, nonetheless, considered structural testing since the specific gates are modeled. Stuck-at faults are tested at each of the inputs and outputs of the gates.

In memories, the cells are tested with a certain set of regular patterns and thus the memory is abstracted in a different fashion than is typically done with logic. Logic is reduced to the gate level, while memory is reduced to functional, electrical, or geometric levels.

When manufacturing type defects are considered from a memory perspective, the defects are examined on a cell or a peripheral circuit basis. A large number of patterns are utilized to test the array of cells and a smaller number of other patterns are used to test the periphery.

3. GENERAL FAULT MODELING

There are numerous fault models that are applicable to memories. While logic predominantly uses the stuck-at fault model plus some other fault models only sparingly, memory fault modeling is much broader.

Defects and fault modeling are discussed in this chapter for all memories. While it would be possible to have a separate chapter for each memory type, it is considered valuable to view them here collectively. Although some of the faults are specific to certain memory types, being aware of memory faults as a whole helps the engineer doing memory fault modeling in the future as new memories or variances of existing memories are developed. As memory designers invent new circuit topologies, new fault models are required and knowing the models for the various memories facilitates that effort.

There are four classic memory fault models. The most classic is the stuck-at fault (SAF) model, which is always used and indicates that a cell is locked in one state or another. A good way to describe these states for defect-free and defective memories is via the use of a Markov diagram [105]. A single cell which is defect free can be written to either state and, when read, retains the information that was originally in the cell. Figure 8-4 shows a Markov diagram of a fault free memory cell. An "R" indicates a read operation while "W0" and "W1" indicate a write "0" and a write "1", respectively. S_0 and S_1 indicate that the cell is in a "0" or a "1" state, respectively. Figure 8-5 shows a Markov diagram of a stuck-at fault in a memory. No matter what happens the cell is in a stuck-at "0" state. Whether it is read or written, regardless of intended data value, the cell stays at a "0".

The next classic fault is a *transition fault model* (TF). In many ways this looks like a stuck-at fault but in this case the memory cell will retain either state. However once it is written to one state it cannot transition back. Thus, when the memory is powered up the cell may be in either a "0" or a "1" state. It can only be written in one direction. The transition fault is illustrated in

Figure 8-6. As shown, it is possible to transition this cell from a "0" state to a "1" state but it cannot transition back.

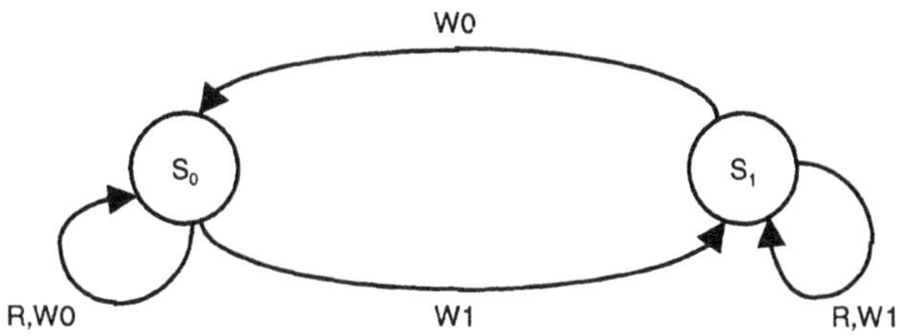

Figure 8-4. Markov diagram for a good memory cell.

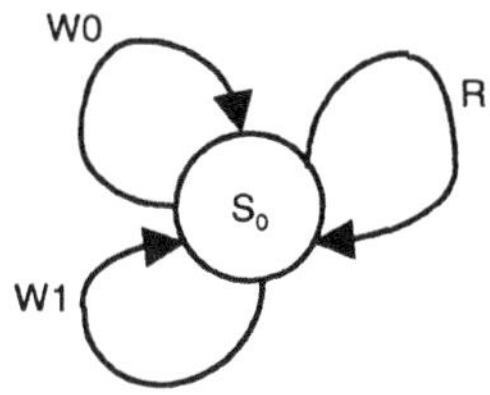

Figure 8-5. Markov diagram for a stuck-at "0" cell.

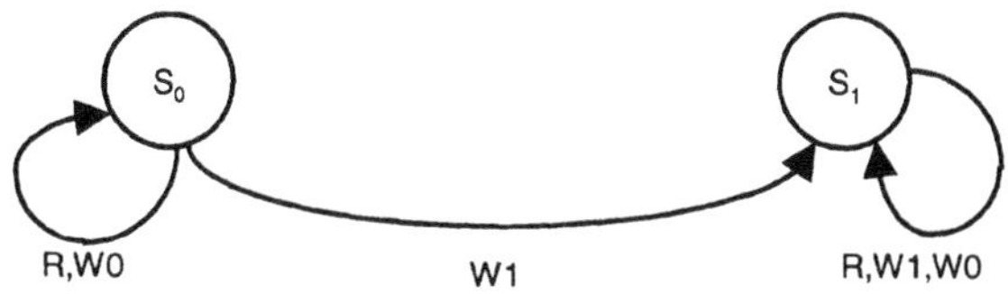

Figure 8-6. Markov diagram of a transition fault model.

The next classic fault is the *coupling fault model* (CF) and there are numerous types of these faults [106]. Simply expressed, a cell can couple into its neighbor cell and cause it to go to an erroneous state or cause it to falsely transition.

Two cells, which are defect free, are illustrated by the Markov diagram shown in Figure 8-7. The cell states and operations are represented by their "i" and "j" subscripts. Each cell can be written to each state. Therefore, there are four possible states, in which the two-cell combination can reside. Each cell can be individually written or individually read. A coupling fault, where a first cell causes a second cell to be written is illustrated in Figure 8-8.

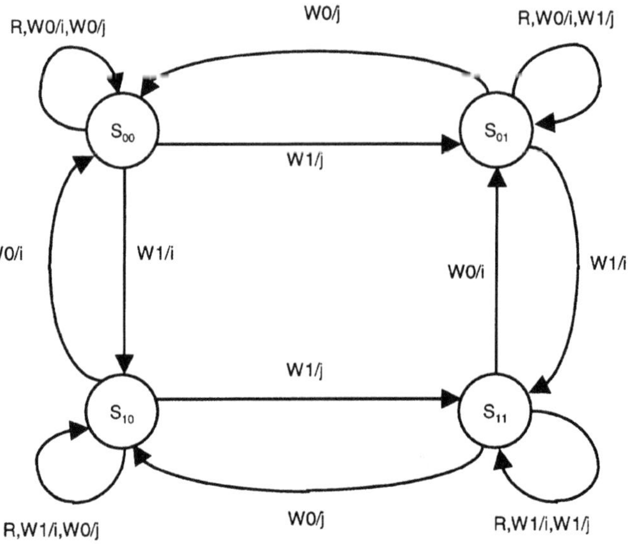

Figure 8-7. Markov diagram of a pair of defect free cells.

It is possible to have a defect where a fault is unidirectional. One cell can couple into another cell but the opposite does not happen. The cell, which does the coupling, is referred to as the aggressor cell and the cell that in turn transitions is called the victim cell. The aggressor cell appears to operate correctly but the victim cell goes to the incorrect state. One way that this type of defect can occur is through a parasitic diode connection between two cells. When a node in the aggressor cell goes high, a node in the victim cell, which is connected by a defect in a diode fashion, is pulled high. When the converse operation occurs, the victim to aggressor diode connection is reversed biased resulting in no transition in the aggressor cell.

It is also possible to have a pair of cells where each affects the other. In this case the fault is bi-directional. A bridging defect can cause such an event to occur.

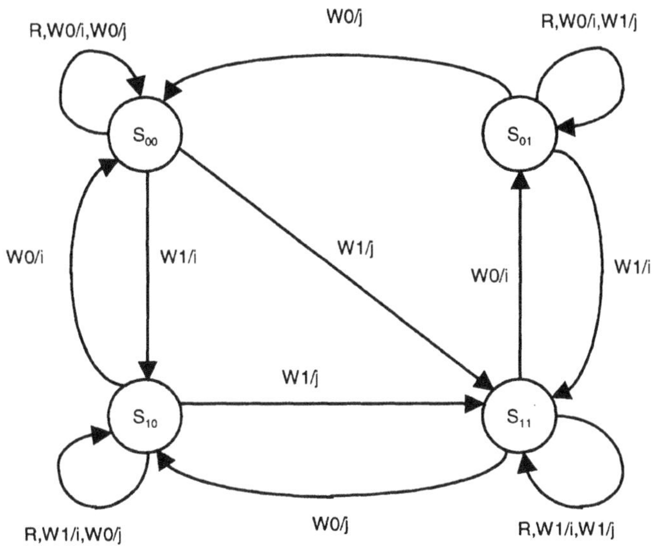

Figure 8-8. Markov diagram of a coupling fault.

The last of the four classic fault models for memories is the *neighborhood pattern sensitive fault model* (NPSF). In this case a memory cell is dependent upon the cells in its neighborhood. Often times memories are described in terms of a nine-cell neighborhood. The base cell in the center is surrounded by eight neighboring cells. The base cell could be dependent on all or a subset of the eight cells around it. The nine-cell neighborhood is described in Figure 8-9. The closest connections are between the base cell and those that are north, south, east, and west but other diagonal interactions are possible. It is also possible for the base cell to be dependent on the other cells in a column or the other cells in a row [107]. Neighborhoods can be defined in various fashions. The neighborhood is considered to include the base cell plus those around it that can affect its behavior. When the base cell is removed from consideration the region is now termed the deleted neighborhood.

Some people have observed that a 25-cell neighborhood dependency is possible [108]. In this case the base cell is in the center and it is possibly dependent upon two cells in each directions.

No further discussion will be pursued on the topic of these classic fault models. There are extensive discussions already in the literature on variations to the classic fault models and coverage here would provide no more insight on the ways that memories should be tested in a manufacturing environment. There are numerous types of faults which are interesting mathematical arrangements. These normally do not, however, reflect real manufacturing type defects. There are other faults which should be examined and which are highly dependent upon the actual circuits that are used. These are the faults that will be discussed in the remainder of the chapter.

Neighbor Cell	Neighbor Cell	Neighbor Cell
Neighbor Cell	**Base Cell**	Neighbor Cell
Neighbor Cell	Neighbor Cell	Neighbor Cell

Figure 8-9. Nine cell neighborhood.

4. READ DISTURB FAULT MODEL

An SRAM has a non-destructive read operation. When data is read from a cell, the data is retained by the cell. There are defects that cause a cell to lose its data during a read. An example of such a defect is shown in Figure 8-10. When a read is performed, charge is pulled off of the bit line and sunk into the cell. The bit line is pulled low through the transfer NFET and the pull-down NFET, numbered T5 and T2 respectively.

When there is no contact to the ground node through the source of T2, the cell will flip on a read operation. A defect like this turns the memory element into a dynamic cell, which temporarily stores a state but which cannot be read. In other words, this defective SRAM loses its data state on a read. This type of defective SRAM should not be confused with DRAM operation. A defect free DRAM loses it data state on a read; there is a write-back operation at the end of a cycle to restore a cell's data after a read. No such write-back operation occurs at the end of an SRAM cycle.

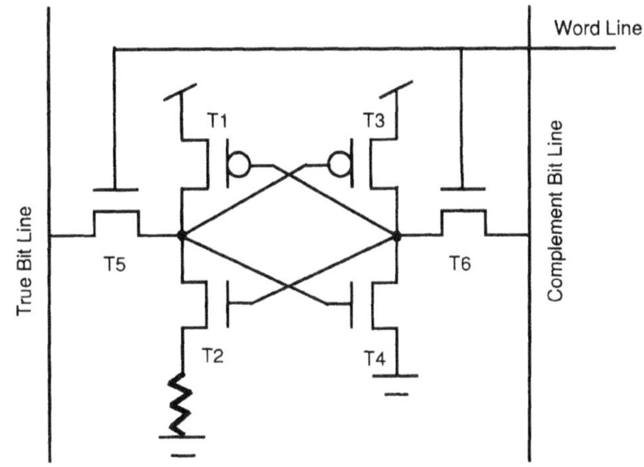

Figure 8-10. Read disturb defect in an SRAM cell.

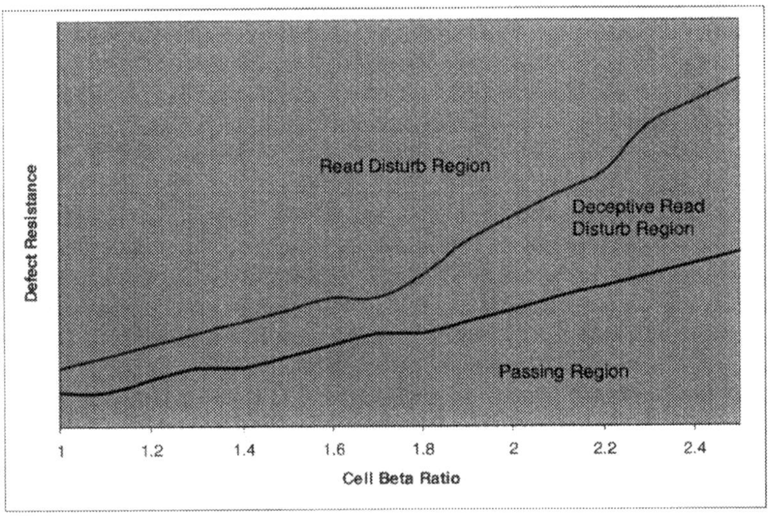

Figure 8-11. Graph of read disturb regions.

In the case of the defective SRAM, if the ground contact is not open but rather is highly resistive, there is a range in which the cell disturbs but the memory's operation does not fail during the first read cycle. It may disturb late in the cycle, with the correct data being placed into the sense amplifier but the incorrect data is retained in the cell. This is referred to as a *deceptive read disturb* or a *deceptive destructive read* [109]. For this fault, there is a certain amount of resistance tolerable in which normal operation can continue to occur. There is a range of resistance that causes the cell to flip immediately, which is then read as the wrong value on the first read. Lastly, there is a region in which the cell disturbs, but only late in the cycle so that it is not detected until the next read operation. These regions are illustrated by the graph in Figure 8-11 where deceptive read disturb region grows as a function of beta ratio.

5. PRE-CHARGE FAULTS

In a memory, it is possible to have a defect which causes the pre-charge circuitry not to operate. One type of a defect is a resistive pre-charge device. Another type of defect is where the pre-charge devices do not turn on due to an open or due to a faulty control circuit for the pre-charge devices. The impact of such a fault is that the bit lines do not pre-charge correctly. Figure 8-12 shows a set of SRAM waveforms where the bit lines do not pre-charge correctly. As a result, one of the bit lines starts out significantly lower than the other bit line. If a "0" is being read, the true bit line should be low. The complement bit line, however, starts out quite a bit lower than Vdd. The result is that the true bit line must discharge for a much longer period of time before it is actually lower than the complement one. Thus, the incorrect value normally is read since the pre-charge circuit does not work correctly.

Since a write forces the bit lines to a full differential potential, it is easier to detect that a pre-charge fault has occurred when a write is followed by a read. When a read is followed by a read, there is less differential that the pre-charge circuit must restore. For a write followed by a read, the write data type needs to be opposite from that of the read to enable detection of this type defect.

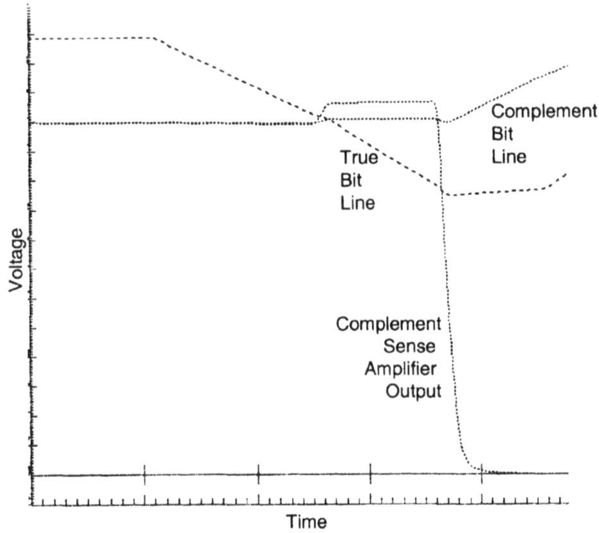

Figure 8-12. Waveforms resulting from a defective pre-charge circuit.

6. FALSE WRITE THROUGH

Another type of fault is the *false write through fault model.* In this case there is a defect on the write head, which causes the data on the input of the write head to be applied to the bit lines and therefore appear on the data output [110,111]. Figure 8-13 shows an example of a write head with a defect that pulls down the true bit line even when the write enable signal is low. It should be noted that the location of a defect can cause either the true or the complement bit line to be pulled low causing an erroneous read.

A second type of defect is shown in Figure 8-14 that can similarly impact a write head circuit. As a result of the possible defect sites, all data type combinations must be utilized in the data being read from the cells and the data input signals being applied to the write head. If a large amount of defect resistance is on the write head then a read operation may barely overcome the incorrectly low bit line. The resulting waveforms look similar to that seen in Figure 8-12, since a bit line is at an incorrect value at the start of a read operation.

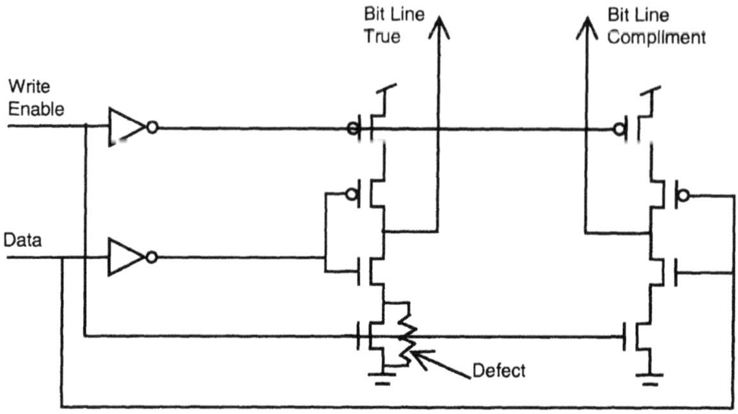

Figure 8-13. Defective write head which imposes signal during a read.

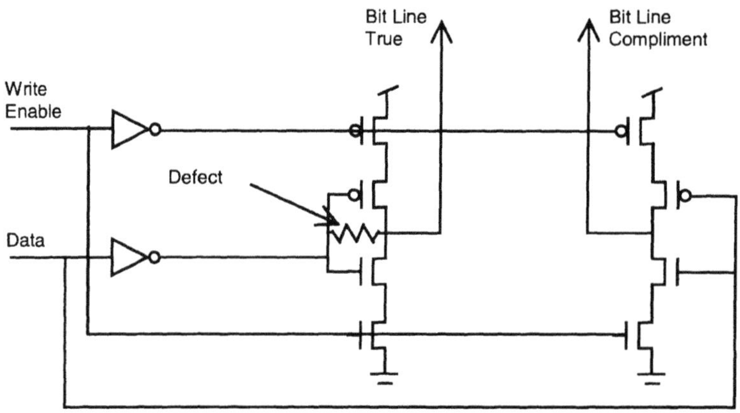

Figure 8-14. Second type of defective write head circuit.

7. DATA RETENTION FAULTS

Another fault model is the data retention fault. Figure 8-15 shows a dynamic RAM cell with a defective resistance discharging the storage capacitor at a faster rate than normal. To detect this type of defect a pause is required to identify that the leakage is greater than the specification of a normal cell. In the case of DRAM cells, leakage off of the storage node is normal. There is an acceptable range of leakage within which the memory operates according to specification. Thus data retention testing on DRAMs really becomes an issue of assuring the specification rather than just testing for manufacturing defects.

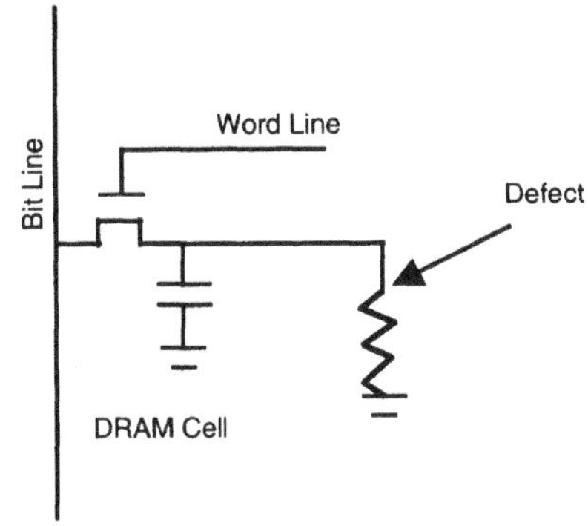

Figure 8-15. A dynamic RAM cell with a data retention fault.

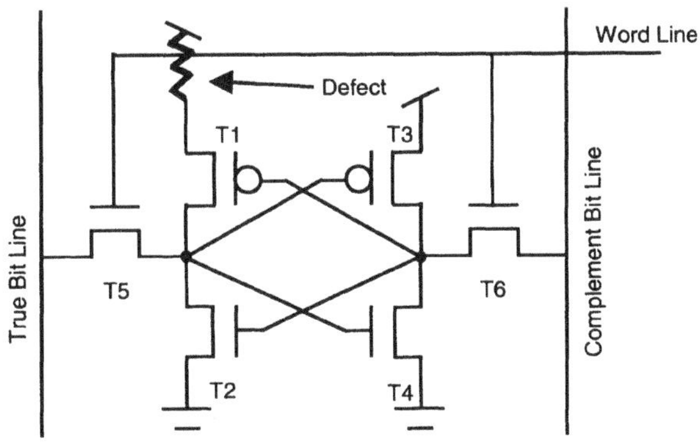

Figure 8-16. Static RAM pull-up type retention defect.

A static RAM cell can also have a data retention fault and there is a variety of possible locations within a cell that can cause data retention issues. Figure 8-10 earlier showed one type of retention fault in the pull-down path. Obviously two sites, one at each of the pull-down NFET sources, can cause a cell to lose its data. Resistive contacts to the drains of these NFETs can similarly cause retention faults. These types of retention faults can be detected by performing a read or possibly multiple reads.

Another type of SRAM retention fault is caused by an open or highly resistive defect in the pull-up path, as illustrated in Figure 8-16. The added resistance can be in either the source or drain of the PFET. A pause, like that of a DRAM retention test, can help find this type of defect. Similarly a bump test or an active pull-down capability, placed on the memory column to facilitate test, can be utilized to find this type defect. More discussion will be included later in the text for ways to find these types of defects.

8. SOI FAULTS

For a silicon-on-insulator technology memory, there are different fault sites as compared with typical bulk silicon cell [112]. Not only can there be shorts and opens along the source, drain, and gate connections but there can be defective connections to a transistor's body node; a specific transistor can have a parasitic or defective connection to the body.

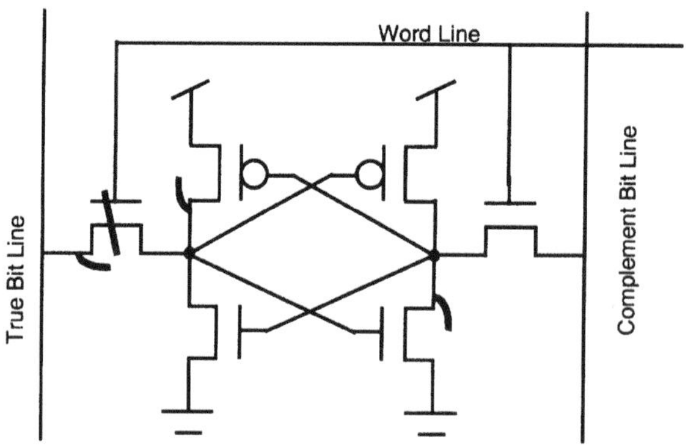

Figure 8-17. A silicon on insulator SRAM cell with defective resistances to body nodes.

Figure 8-17 shows four possible defective connections between the bodies of FETs in an SRAM cell and nearby nodes that can destabilize a cell that is intended to store a "0". With a body being brought high, as the result of a defective resistance to another node, a FET may be more powerful than the design intended. If this happens on a transfer FET, more charge can be pulled out of the cell's low node and the cell can flip. Similarly, a pull-down device may be stronger than intended causing a node that was supposed to be high to be brought low, again destabilizing the cell. Other similar SOI defects may exist that destabilize the cell from different physical locations but which have the same impact for each data type.

9. DECODER FAULTS

A key element of random access memories is the ability to access the various storage locations in any order that is desired. To accomplish this, a decoder circuit takes an address supplied to the memory and decodes a specific location. Selecting a specific row means that the appropriate word line is brought high. Selecting a specific column means that the appropriate bit line is steered through a multiplexer to the sense amplifier circuit. The logic that performs this decode operation can be typical NAND or NOR gates. These gates, if in random logic, would be well tested by automatic test pattern generation (ATPG) software. The gates, in the context of a memory, must undergo more rigorous testing. The outputs of the decoder are not observable and so cannot be tested with logic patterns. The decoder is tested to ensure that the appropriate location is accessed.

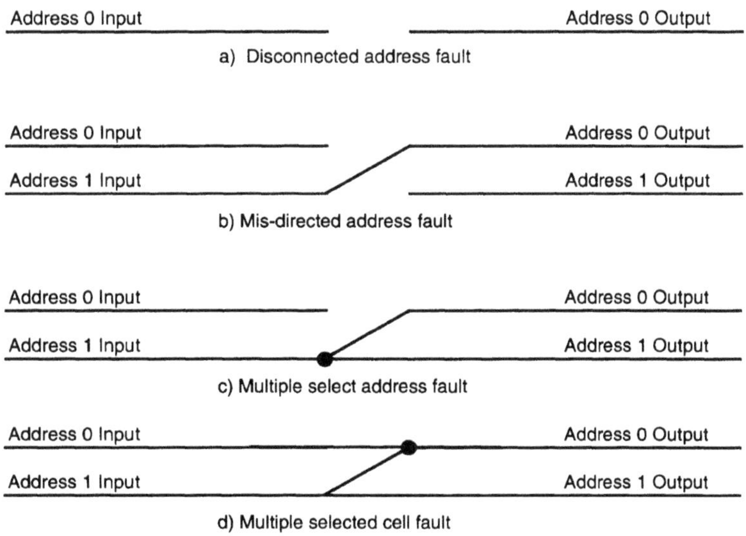

Figure 8-18. Four decoder fault possibilities.

It is possible to have a fault where no location is accessed, the wrong location is accessed, more than one location is accessed, or multiple addresses access a single location [113,114]. Figure 8-18 shows these possible decoder or address faults (AF). No address could be accessed due to a stuck-at fault in the logic or due to an open in the path to the memory element. The wrong location could be accessed due to a stuck-at fault. Multiple addresses could be accessed or multiple cells could be selected, although less likely, due to a bridging short in the path to the memory elements.

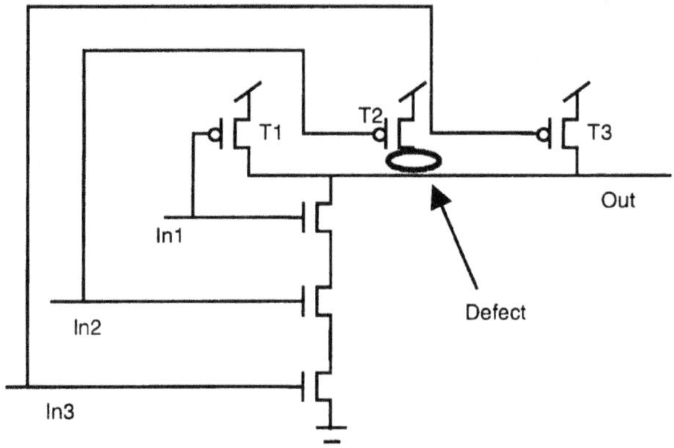

Figure 8-19. Static decoder open fault.

Static decoders have specific faults of their own. Figures 2-22 and 2-23 showed the difference in chapter two between the design of a static decoder and that of a dynamic decoder. If one considers the static decoder, which can be illustrated by a simple NAND gate, the possibility for faults can be observed. If there are more than two inputs to a static decoder then a *static decoder open fault* can occur. A NAND gate is shown in Figure 8-19 with an open on a PFET that can easily be missed during test [115,116]. When sequentially incrementing or decrementing through an address space the NAND gate will appear to operate correctly. First T1 is on and then during the following cycle T2 should keep the output high. Since this latter PFET is defective and cannot pull the output high, one would think that this defect could then be detected. In fact the output remains high due to capacitance on the node thereby masking the defect from detection. Similarly, when decrementing through the address space, T3 is on and then in the following cycle the defective PFET T2 should hold the output high. Instead the output is again held high for capacitive reasons. Tables 8-2 and 8-3 show the result of incrementing and decrementing through the address space. The "~1" notation indicates that the node is a "1" for capacitive reasons only and is not being actively held, as it should be in a defect-free case.

To detect this type of defect the output must be set up to transition from low to high as a function of each pull-up PFET. A NOR gate has the same susceptibility to static decoder opens however the possible opens are in the NFET pull-down path instead of the PFET pull-up path. The same type of defects can impact a dynamic decoder but they are easily detected and require no special testing. Since a dynamic decoder pre-charges in one direction and then evaluates in the other direction each cycle, an output node cannot be erroneously held at a value due to capacitive mechanisms [117]. Simple incrementing and decrementing through an address space identifies these defects. A pattern which facilitates testing for static decoder opens can also test for *slow-to-decode transitions* and therefore may be warranted even when static decoders are not employed in the design of a memory.

Table 8-2. Incrementing addresses with a static decoder open defect.

	In2	In1	In0	Out
1	0	0	0	1
2	0	0	1	1
3	0	1	0	1
4	0	1	1	1
5	1	0	0	1
6	1	0	1	~1
7	1	1	0	1
8	1	1	1	0

Table 8-3. Decrementing addresses with a static decoder open defect.

	In2	In1	In0	Out
1	1	1	1	0
2	1	1	0	1
3	1	0	1	~1
4	1	0	0	1
5	0	1	1	1
6	0	1	0	1
7	0	0	1	1
8	0	0	0	1

10. MULTI-PORT MEMORY FAULTS

A multi-port memory has multiple paths into and out of a memory. These paths can interact in a defective memory in curious ways. It is important to understand the possible defective operations to ensure that multi-port memories are adequately tested and that the resulting chips are indeed good.

There can be faults that cause improper interactions between different addresses but that are on the same port. These are referred to as intra-port defects. An example is given in Figure 8-20 where adjacent metal word lines are shorted. In this case two addresses can be accessed at the same time incorrectly due to a short between the word lines. Both of these word lines are for port B and therefore the defect is an *intra-port fault* [118].

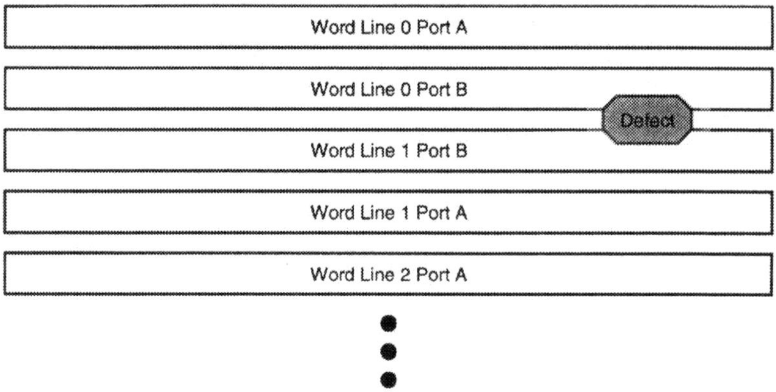

Figure 8-20. Intra-port word line defect.

A defect which can cause improper operation between two different ports is referred to as an *inter-port fault* [119]. Figure 8-21 shows a short between adjacent metal word lines where two different ports on the same address are connected by a defect. This type of defect causes an inter-port intra-address fault. An inter-port inter-address fault is also possible between shorted metal bit lines.

When examining multi-port memory faults it is important to understand that different operations have different impacts [120]. For example, a write operation dominates a read operation. Since a write involves a full bit line swing and a read involves only a small swing on a bit line, a read is less severe. Given that the objective is to identify defects, a more sensitive operation facilitates fault detection. Since more than a single operation can go on simultaneously in a multi-port memory, a more sensitive operation used in tandem with a more severe operation will aid identification of erroneous interactions. The more severe operation should be used with the aggressor port and the more sensitive operation should be used with the victim port. A two-port memory example of this is shown in Figure 8-22 where a write "0" on port A is occurring on the 0th word line while a read "1" on port B is occurring on the 63rd word line. The anticipated defect-free resulting bit-line potentials are shown on the bottom of the figure. Since there is a short between the 0th word lines, the write will also cause a read to occur on port B. The read on the 0th and 63rd word lines will conflict, causing both the true and complement bit lines to discharge, rather than just the true bit line. As a result, the sense amplifier will set randomly, causing the defect to be detected after a number of reads have occurred. The diagram shown in Figure 8-22 illustrates schematically the defect shown from a layout point of view in Figure 8-21.

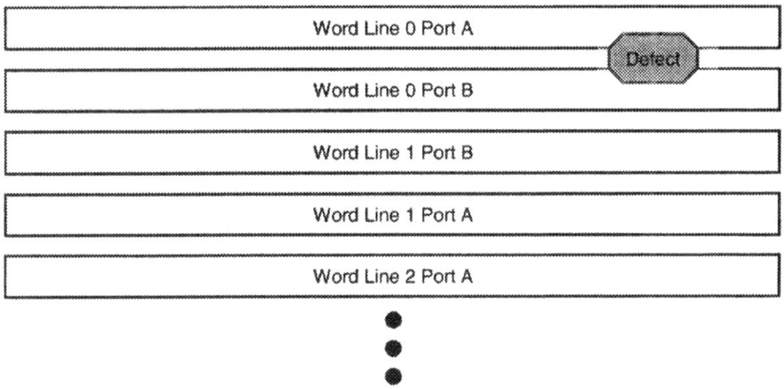

Figure 8-21. Inter-port bit line defect.

Obviously as higher dimension multi-port memories are utilized more complex interactions are possible. It is important to understand these possible interactions, based on possible real defects, and test the memory accordingly.

Using a write in conjunction with a read, and understanding relative adjacencies, helps test for shorts between bit lines. Figure 8-23 shows a group of bit lines where the port A bit lines are reading a "0" and the port B bit lines are writing a "1" along with the defect-free potentials. During the write, the complement B port bit line is driven to zero volts while the true bit line is maintained high. During the read, the true port A bit line discharges by a small amount, denoted here by "~100 mV". Since there is a short between the complement B port bit line and the complement A port bit line, the complement A port bit line is driven lower in potential than the true A port bit line. As a result, the sense amplifier detects the erroneous differential on the A port bit lines since the actual operation does not match the intended operation. It is key to have the correct adjacency and data type to test for these multi-port memory defects. If the B port bit lines were mirrored, writing a "0" on the B port while reading a "0" on the A port would have facilitated detecting a bit line short. When only two ports exist, all possible combinations can be used.

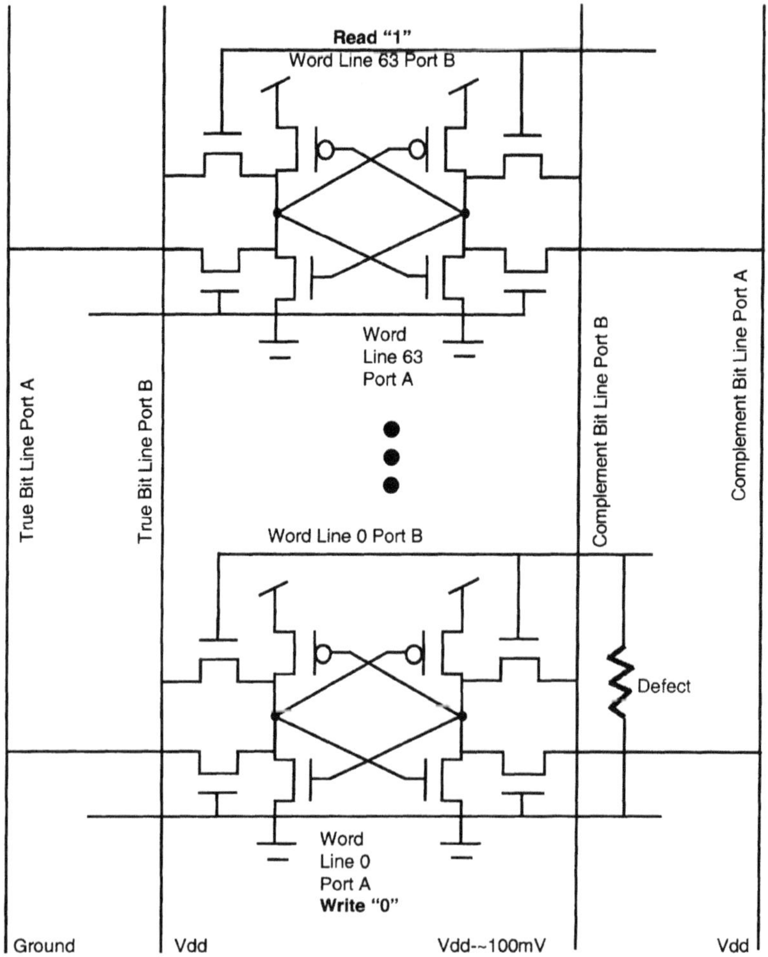

Figure 8-22. Simultaneous read and write operations with a defect present.

When a higher dimension multi-port memory exists, performing all possible data type combinations in concert with all possible operation combinations becomes problematic. For higher dimension multi-port memories, purposeful understanding of adjacencies, operation impacts, and possible fault sites must be utilized to select the most beneficial patterns and thereby achieve the desired quality results. In addition, certain multi-port memories have ports that only have one operation, such as a read only port. In this case it is not possible to perform a write on this port and look for interaction with an adjacent port's bit lines but neither can a write happen during functional operation. There is, however, the reduced ability to detect a short to an adjacent bit line. Without making modifications to the memory, to include design-for-test features, the best test that can be provided is to perform a read on adjacent bit lines with opposite potentials that looks for shorts which could cause an interaction failure. If no failure is detected during testing, there is little chance that this part could fail in the field.

11. OTHER FAULT MODELS

There are numerous other fault models discussed in the literature. Extensive research has been done to identify various ways that memories fail and in the ways that they can be mathematically represented to fail. In appendix A more fault models are reviewed. They are included for reference sake and are not necessarily critical to one's understanding of real memory fails and the required memory manufacturing testing. They can, however, be helpful for understanding the large body of information on memory testing and the associated patterns.

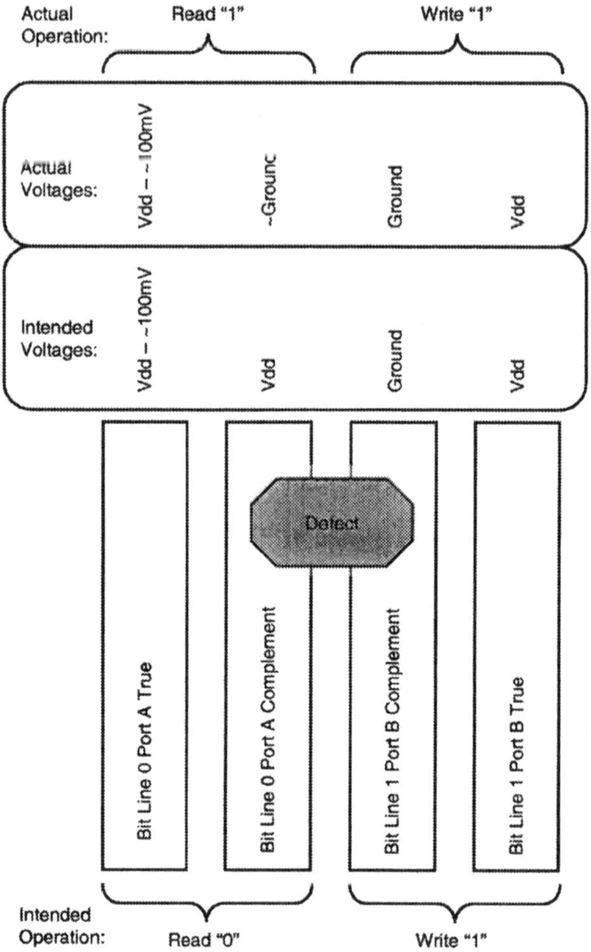

Figure 8-23. Bit line short causing an inter-address inter-port fault.

Chapter 9

Memory Patterns

Testing Issues

"A friendly eye could never see such faults." – from Shakespeare's *Julius Caesar*

Patterns are the essence of memory testing. Memories have many analog circuits and since the memory circuits are packed tighter than anything else on the chip, special patterns are required. These patterns look for weakness in the analog circuitry and for interaction between the tightly packed adjacent structures. Memories are regular structures, requiring extensive regular patterns to facilitate testing. If a poor set of patterns is utilized, memories will pass test when they are in fact defective. It is not unusual to hear someone state that they are using the "such and such pattern" and that it "covers everything." This type of statement is clearly ignorance. The ignorance involves not understanding the memory design, the possible fault models, and the capabilities of the respective patterns. No single pattern is sufficient to test a memory for all defect types [121]. A suite of patterns is required to catch and eliminate the real defects that can happen in a manufacturing environment.

Many people incorrectly approach memory testing with a logic mentality. A memory stores ones and zeros. If zeros are written to all addresses and read from all addresses, half of the defects would be covered, right? Then once ones are written and read from all addresses the other half are covered, right? Wrong.

1. ZERO-ONE PATTERN

The pattern just described is called the Zero-One pattern. It is described in Table 9-1 and the simplicity of the pattern becomes obvious. Some refer to this pattern as the blanket pattern [122] or as MSCAN. Each line in the table represents a full address sweep. The " ⇕ " denotes that the order is non-consequential and that no preference is given to the addressing order. Each address is accessed only four times and is referred to as a 4N pattern. This nomenclature is used regularly when talking about pattern length. A 4N pattern accesses each location four times, a 9N pattern accesses each location nine times, and so on. If a memory was only sensitive to stuck-at faults, the Zero-One pattern would be sufficient to catch all defects. With fault grading, the Zero-One pattern achieves 100% stuck-at coverage in the memory cells. Toggle coverage would again be 100%. (Toggle coverage tracks the internal nodes of a circuit to see the percentage of the nodes which are set to each state. Some circuits which, may be difficult to fault grade, can be examined for the toggle coverage. Toggle coverage is imprecise, since it doesn't tell the exact coverage to specific faults, but provides qualitative insight into the amount of the circuit which is being exercised.)

Table 9-1. Zero-One pattern description.

1	W0 ⇕
2	R0 ⇕
3	W1 ⇕
4	R1 ⇕

Since the Zero-One pattern has 100% stuck-at coverage and 100% toggle coverage of the memory cells, it could be thought that this pattern alone would be sufficient for memory testing. This is a gross fallacy and the Zero-One pattern should never be applied as a sufficient test for any memory. The Zero-One pattern does not provide coverage for data retention, deceptive destructive read, address decoder, un-restored write, and numerous other fault models. Some people cannot comprehend the need for a book or even a paper on memory testing, since their concept of a memory test is completed with the Zero-One pattern. The Zero-One pattern does not indicate whether each cell can be uniquely addressed. In fact, it would be possible for a memory to pass test to the Zero-One pattern if only a single cell worked in the memory. Figure 9-1 shows an intended eight by four memory where, no matter which address is input, only one address can be selected. Further, for all of the data inputs and data outputs, all of the data bits could be fanned into and fanned out of a single cell. Clearly, this memory is grossly defective and should never have passed test. It would, however, pass the Zero-One test, thus demonstrating the gross unacceptability of this pattern.

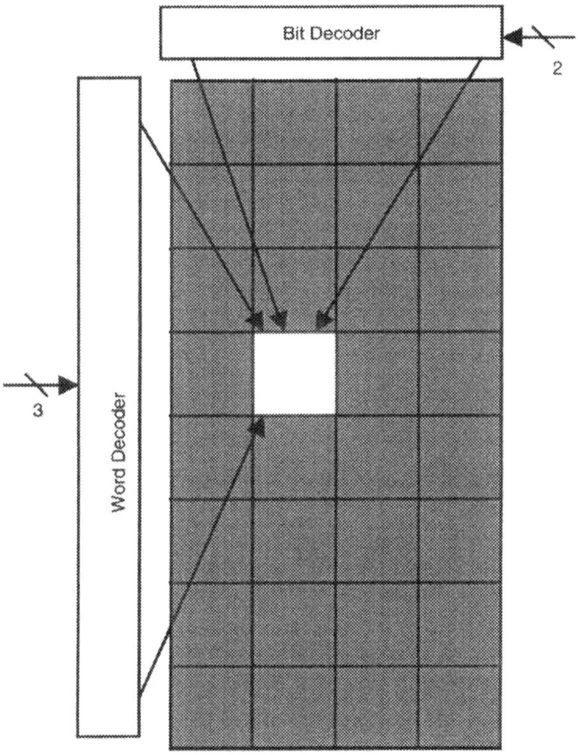

Figure 9-1. Faulty memory which accesses only one cell but passes the Zero-One pattern.

2. EXHAUSTIVE TEST PATTERN

It is obvious that a simple pattern is insufficient to test any memory. A more complex pattern must be utilized, but which pattern and how should it be determined? One possibility is to utilize an exhaustive pattern. An exhaustive pattern is one in which all possible data combinations are included. For an "n" input combinational circuit, the number of patterns required is 2^n tests to hit all possible combinations. When memory elements are introduced there are 2^m required patterns, where there are "m" memory elements, to achieve exhaustive testing. The result is that 2^{m+n} patterns are required to execute an exhaustive pattern on a sequential (a circuit which contains some memory elements) circuit.

A memory circuit is far more than a sequential circuit, however. A memory is dominated by its memory cells. An exhaustive pattern is considered to be all of the possible combinations of the memory cells themselves. The concern is about faulty interaction between the memory cells. If the possible combinations are covered for the memory cells, the inputs are thoroughly exercised. If there are "n" memory cells then clearly 2^n combinations are possible. But for each cell, the original state must be read, it must be written to the opposite state, and then be read to verify that state. This provides a 3n multiplication factor. Thus an exhaustive pattern requires $3n2^n$ tests. A $3n2^n$ pattern requires much more time than is feasible. For example, a small 32 K byte cache, exercised at 100 MHz requires $10^{78,903}$ years to complete an exhaustive test. Clearly an exhaustive test is not a viable option.

Since a simple test cannot suffice and an exhaustive test is impossible to implement due to time limitations, another pattern must be selected. This pattern selection needs to be done based on an intelligent understanding of the memory design and its associated fault models. A number of the key memory test patterns will now be described along with some information on the faults which they covered. There is discussion in appendix B of additional memory test patterns which are used or referred to in the related literature and therefore can be helpful in understanding.

Key point: Test patterns must be selected based on the memory topology.

3. WALKING, MARCHING, AND GALLOPING

There are a number of terms which are helpful in understanding memory test nomenclature. Patterns are often broadly referred to in the categories of walking, marching, and galloping patterns [123].

A walking pattern has a single cell or a single address, which is in a different state from the other locations in the memory. A single cell is considered only when a bit-oriented memory is under review. Since word-oriented memories are the norm, they are considered here and the unique location is therefore an address that accesses a word's worth of bits at a time. A walking pattern has a blanket background entirely of one data type. There is a single address with a different data type on any give cycle. If a blanket background of zeros exists and a one is being "walked" through the memory, then only one address will be in the "1" state. All of the addresses covered before the current cycle are a "0" and all of the locations yet to be covered are in a "0" state. Each address could have the sequence "read zero, write one, read one, write zero" applied to it. This changes the state of a specific address during test but returns the data to the state of the blanket background before continuing on to the next address. The sequence is described in Table 9-2, with the " ⇑ " indicating that the addresses are sequenced incrementally from the all zeros address through the maximum address in the memory space. A comma separates operations that occur on successive cycles. All four operations are performed on each address before proceeding to the next address.

Table 9-2. Walking pattern element example.

1	R0, W1, R1, W0 ⇑

A marching pattern involves changing the data state at a given address and leaving the address in the changed state when proceeding to the next memory address. Before a march sequence is begun, a background data type exists in the memory. When the march sequence is half way through the memory space, half of the memory will be in the old data type and half will be in the new data type, which is being placed there by the march sequence. At the end of the march sequence all of the memory will be in the new data type. Table 9-3 provides an example of a march pattern element. This pattern assumes that a blanket zero pattern already exists in the memory. At each address location zeros are read from the memory, ones are written into it, and the ones are read. The address space is being decremented through, as indicated by the " ⇓ " in the table.

Table 9-3. Marching pattern element example.

1	R0, W1, R1 ⇓

While walking and marching patterns are primarily data oriented pattern descriptions, a *galloping* pattern is primarily an address oriented pattern description. A galloping pattern provides a ping-pong action between the base cell under test and each other cell in the memory. Since every address must be accessed, as the base location is incremented to a new location, the test time is very long. This pattern requires $2(N+2n^2)$ cycles to complete. Whenever a pattern includes an "n^2" term the number of cycles is simply too long to be practical in a manufacturing environment [124]. The fault coverage of a galloping pattern is exceedingly high but the test duration relegates its use to characterization purposes and not those of manufacturing test.

An alternative to a full galloping pattern is to isolate the galloping within a row or within a column. The number of overall cycles drops since there is no longer an "n^2" term that covers the entire array. For a square memory array the total number of cycles becomes 2(N+2n+2n*Sqrt(n)). The ping-pong iteration now remains within a row or within a column but the resulting interaction test is extensive.

4. BIT AND WORD ORIENTATION

The terms bit-oriented and word-oriented have already been mentioned but due to their preponderance in the literature it is worth spending some time clarifying these terms. Early memories might have only a single bit data input and data output, thus the term bit-oriented. In this case each memory cell could be addressed individually. Subsequent to those early memories were memories which had wider data input and data output buses. Each time a write or a read operation was performed, the full width of the data bus was utilized. These latter memories are referred to as word oriented. Patterns have been developed over the years assuming a bit oriented memory and these patterns have been modified to accommodate the word orientation normally utilized.

Table 9-4. Data backgrounds and their inverses utilized for an eight-bit word.

D7	D6	D5	D4	D3	D2	D1	D0
0	0	0	0	0	0	0	0
0	1	0	1	0	1	0	1
0	0	1	1	0	0	1	1
0	0	0	0	1	1	1	1
1	1	1	1	1	1	1	1
1	0	1	0	1	0	1	0
1	1	0	0	1	1	0	0
1	1	1	1	0	0	0	0

Since each memory cell can no longer be addressed in a word-oriented memory, varying data background patterns have been generated to ensure that cell interactions are covered by the tests. Where "m" is the number of bits in a word, the number of data background patterns required was $\log_2(m)+1$. A memory with an eight-bit word thus requires 3+1, or 4, data background patterns [125]. It is always assumed that a background and its inverse are to be utilized. Table 9-4 shows the various data backgrounds utilized along with their inverses for an eight-bit word example. As the number of bits in a memory's word grows, the number of background patterns grows as well. It is not unusual to have 128 bit or larger memory interfaces for embedded memories and thus a large number of background patterns can be generated. The data background patterns for a 64-bit word are included in appendix B.

Based on Figure 2-28, it can be seen that memories normally have the bits in a word spread across multiple sub-arrays. Since these bits are widely separated from one another there is little opportunity for them to interact with each other. For this reason each sub-array can be treated as a separate bit-oriented memory and the patterns can reflect this in their data background patterns. If the memory word is not spread across sub-arrays then the bit interaction and multiple data backgrounds must be required. These factors again point out the need for understanding the details of the memory design to provide high quality and efficient memory testing.

The backgrounds discussed thus far in this section are covering interactions for bits within a word. There can be varying data background test requirements based on the cell adjacencies. For example, in SRAMs there may be a requirement to execute a physical checkerboard pattern in the memory. This pattern checks for differing leakage paths from blanket patterns between a base cell and its immediate neighbors. For DRAMs, a richer data pattern is often required. These varying physical background patterns do not necessarily require placing the data inputs in a word into a all possible of permutations. To achieve a physical checkerboard pattern, all of the data inputs can be fanned out from a single test data input. As the addresses are stepped through to access all of the locations in the memory, the test data input needs to be set at the correct binary value to facilitate obtaining the result of a physical checkerboard within each sub-array. Thus independent control of each data input is not required during test. This fact greatly eases the challenge of obtaining an efficient built-in self-test and will be covered in later chapters.

5. COMMON ARRAY PATTERNS

Many times a physical configuration of memory cell data types is desired to enable detection of various defects. The simplest of the physical configurations is a *blanket pattern* where the memory is either in an all ones or an all zeros state. Whenever a physical configuration is chosen, both the true and the complement data patterns need to be applied to the memory. Thus a blanket pattern requires, at some point, to have the all zeros state applied and at another point to have the all ones state applied. While intuition may indicate that the blanket pattern is not severe from a stress point of view, this perspective is not warranted. Since most memory cells, the associated bit lines, and the respective sense amplifier circuitry have both a true and a complement portion the voltage potentials of adjacent structures must be considered. If the true and complement portions of a cell are represented by "T" and "C" respectively, then two cells which are just stepped horizontally could be described by the nomenclature "TCTC". In this case a "0" that is stored in both cells would drive potentials reflecting a "0101" in the "TCTC" portions of the cell. Thus the adjacent structures would each be in opposite voltage potentials, even though the cells are both in the same state. If the two cells are mirrored about the "y" axis then "TCCT" would be the describing nomenclature. In this case a blanket zero pattern would produce a "0110" in the "TCCT" structure. With this type of mirroring, a blanket pattern does not produce optimal stress conditions. Chapter two can be examined for a more detailed discussion of physical adjacencies.

0	1	0	1
1	0	1	0
0	1	0	1
1	0	1	0
0	1	0	1
1	0	1	0
0	1	0	1
1	0	1	0

Figure 9-2. Physical checkerboard in memory array.

Figure 9-2 shows the next array pattern, which is the *physical checkerboard.* It is simply an alternating sequence of ones and zeros in both the "x" and "y" dimension as the name implies. It should be noted that a physical checkerboard is not necessarily obtained by applying a logical checkerboard pattern. Applying a logical checkerboard on occasion generates a physical blanket pattern that alternates on a sub-array basis. A physical checkerboard can be quite useful in looking for leakage between adjacent cells. Through a combination of the blanket pattern and the physical checkerboard pattern, worst case testing is achieved for many memory topologies.

A row stripe pattern can be helpful in looking for problems between adjacent rows [126,127]. A row stripe, as illustrated in Figure 9-3, is often referred to as a word line stripe pattern.

0	0	0	0
1	1	1	1
0	0	0	0
1	1	1	1
0	0	0	0
1	1	1	1
0	0	0	0
1	1	1	1

Figure 9-3. Row stripe in memory array.

0	1	0	1
0	1	0	1
0	1	0	1
0	1	0	1
0	1	0	1
0	1	0	1
0	1	0	1
0	1	0	1

Figure 9-4. Column stripe array pattern.

A column stripe pattern can be useful in examining interactions between adjacent cells horizontally and between adjacent bit lines. This pattern is sometimes referred to as a bit line stripe pattern. Figure 9-4 displays a column stripe pattern.

6. COMMON MARCH PATTERNS

6.1 March C- Pattern

The discussion thus far in the chapter has covered types of patterns and some of the relevant nomenclature. At this point specific patterns will be addressed with their sequence of operations and covered fault models. The reader should examine appendix B for details on the patterns not covered in this chapter. The patterns in appendix B are considered to have certain inefficiencies, be out of the mainstream of memory test, or not be crucial to memory test, given the existence of other memory test patterns. The patterns described in this chapter are those widely utilized in the industry, commonly referred to in the literature, or which have significant benefit in finding manufacturing defects.

The March C- pattern is a slight reduction of the March C pattern, which had an inefficiency or redundancy that was eliminated by the March C- pattern. (The March C pattern is covered in appendix B along with the March A, March B, and numerous other patterns.) The March C- pattern is described in Table 9-5. Each element of the March C- pattern is shown on a successive line. An element describes a sweep through the full memory space. The choice of utilizing a separate line for each element is maintained throughout the text. An alternative representation, which is well known in the literature, is shown in Table 9-6. While this latter representation clearly describes the pattern sequence and requires fewer column inches of space, it has its limitations as one looks at more complex patterns. As descriptions of multi-port memory patterns are generated, separating one element from another by semi-colons can become cumbersome.

Table 9-5. Description of the March C- pattern.

1	W0 ⇕
2	R0, W1 ⇑
3	R1, W0 ⇑
4	R0, W1 ⇓
5	R1, W0 ⇓
6	R0 ⇕

Table 9-6. Alternative description of the March C- pattern.

W0 ⇕ ; R0, W1 ⇑ ; R1, W0 ⇑ ; R0, W1 ⇓ ; R1, W0 ⇓ ; R0 ⇕

The March C- pattern is very simple and yet it powerfully identifies a huge number of real defects seen in manufacturing test. The March C- pattern is ostensibly stated as covering unlinked idempotent coupling faults.

In reality it covers a host of faults including stuck-at faults, many coupling faults, transition faults, and the like. Many decoder faults are also covered, in that the March C- pattern is the simplest of the "unique address" patterns. A unique address pattern is one which checks for the four simple decoder faults described in Figure 8-18. If the address decoder points to the wrong address, to multiple addresses, or to no address on any given write in the March C- pattern, then a subsequent read detects the failure.

As an example, suppose that multiple addresses are selected due to a defect in the decoder. Suppose that when a write "1" is performed on word line five that word line three is selected as well and a write "1" is performed there also. The first through third elements of the March C- pattern can successfully complete. The fourth element activates the defective operation. During the address decrement of element four, word line five is reached and "1" is written. Due to the defect, word line three is also written to a "1". As element four continues its decrementing sweep through the array, when word line three is reached a "1" is read whereas a "0" is expected. The defective operation is now identified.

The March C- pattern is considered a 10n pattern, due to its ten operations, two for each of elements two through five and one for each of elements one and six. Again, the small "n" denotes the number of operations on each individual cell for a bit oriented memory. Since the multiple bits are normally divided across the multiple sub-arrays for most memories, the March C- pattern can be used as a 10N pattern with no loss of defect detection. The "N" denotes the number operations on each address, rather than each bit, within the memory.

6.2 Partial Moving Inversion Pattern

The March C- pattern is considered a unique address pattern and specifically can be considered a two-step unique address pattern, since each of its elements two through five constitute two operations. The Partial Moving Inversion (PMOVI) pattern is the next in a series of logically progressive patterns and can be termed a three-step unique address pattern. A unique address pattern facilitates ensuring that each address is uniquely addressed in the proper fashion. The most common address decoder problems are detected by unique address patterns. The PMOVI pattern is very similar to the March C- pattern but utilizes three operations in elements two through five. These can be seen in Table 9-7.

By reading each cell immediately after it is written, as is done in the third operation of this three-step unique address pattern, the cell is shown to have been correctly written. This prevents a defective operation from later writing a cell which had failed to be written correctly. In this case the later writing of a defective cell can mask the incorrect operation and thus avoid detection. The third operation of reading immediately after being written prevents such masking and, in the event that a cell is destabilized during a write, reading the cell immediately after the write allows detection. The read of the unstable cell occurs prior to its being able to recover with the correct data value. The Partial Moving Inversion pattern is a 13N pattern.

Table 9-7. Description of the Partial Moving Inversion (PMOVI) pattern.

1	W0 ⇓
2	R0, W1, R1 ⇑
3	R1, W0, R0 ⇑
4	R0, W1, R1 ⇓
5	R1, W0, R0 ⇓

6.3 Enhanced March C- Pattern

The next in the series of unique address patterns is the Enhanced March C- pattern. It is similar to the March C- and Partial Moving Inversion patterns but utilizes four operations in each of elements two through five. It has also been referred to as the Unique Address Ripple Word pattern, when it is performed modifying row addresses most frequently. This pattern will catch all of the defects identified by the March C- and Partial Moving Inversion patterns but will also detect pre-charge defects.

Detecting pre-charge defects is accomplished by the rapid succession of the fourth operation in an element being performed immediately prior to the first operation of the same element, at the next successive address. Thus for element two in Table 9-8, a write of a "1" at word line 11 is followed immediately by a read of a "0" at word line 12. A defect can prevent the bit lines from pre-charging correctly. The write of a "1" requires the true bit line to be held at Vdd and the complement bit line to be held at ground. The following read of a "0" requires the complement bit line to fully be restored to Vdd, while the true bit line only discharges a small amount. If the complement bit line does not fully restore to Vdd, incorrect data value will be read on the subsequent read of a "0". The Enhanced March C- sequence thus allows detection of subtle pre-charge defects.

The addressing order is critical to detection of the pre-charge defects. The same column must be utilized while successively addressing different rows. Thus the rows must be rippled most frequently. The four-step sequence is sometimes utilized while rippling columns most frequently but the benefit is not as clear.

The Enhanced March C- is considered by the author as the starting point for most memory test pattern development. If a memory does not use the same bit lines for reading and writing the memory then the Partial Moving Inversion pattern is considered the appropriate starting point. The Enhanced March C- pattern requires 18N to complete.

Table 9-8. Description of the Enhanced March C- pattern.

1	W0 ⇕
2	R0, W1, R1, W1 ⇑
3	R1, W0, R0, W0 ⇑
4	R0, W1, R1, W1 ⇓
5	R1, W0, R0, W0 ⇓
6	R0 ⇕

6.4 March LR Pattern

The next pattern that is often used in industry during memory testing is the March LR pattern [128]. It is really a combination of marching and walking elements, as can be seen in Table 9-9 and was developed to detect realistic linked faults. The first two elements match that of the March C- pattern. The third element is a walking sequence as at any given step there is only one address which contains 0s. Similarly, the fifth element is a walking sequence where only the base cell contains 1s. This is a 14N pattern which combines the marching aspects included in the March C- pattern coupled with a walking zero and walking one element.

Table 9-9. Description of the March LR pattern.

1	W0 ⇕
2	R0, W1 ⇓
3	R1, W0, R0, W1 ⇑
4	R1, W0 ⇑
5	R0, W1, R1, W0 ⇑
6	R0 ⇑

6.5 March G Pattern

The March G pattern is shown in Figure 9-10. This pattern includes a pause in the sequence to facilitate retention testing [129,130]. The last two elements include the three-step unique address sequence. This pattern is useful for finding stuck open defects. It is a 23n pattern.

Table 9-10. Description of the March G pattern.

1	W0 ⇕
2	R0, W1, R1, W0, R0, W1 ⇑
3	R1, W0, R1 ⇑
4	R1, W0, W1, W0 ⇓
5	R0, W1, W0 ⇓
6	Pause
7	R0, W1, R1 ⇕
8	Pause
9	R1, W0, R0 ⇕

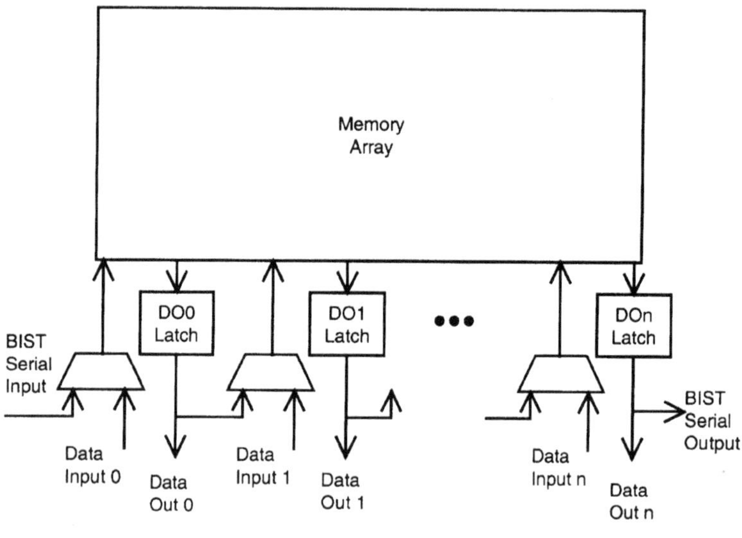

Figure 9-5. Data output to data input connections for an SMarch self test.

7. SMARCH PATTERN

The serial march or SMarch pattern utilizes one data output from a memory bit as the next data input value to be written [131]. In this manner the memory cells are serially modified through successive reads and writes to the same word. Coupling faults are detected, even if the bits within a word are immediately adjacent to one another. If the data inputs and data outputs are on opposite sides of the memory, a wiring problem exists to get each data output connected to the next data input during self test. Figure 9-5 shows a configuration connecting the data outputs and data inputs as required for the SMarch pattern. Pseudo-code describing the SMARCH pattern is included in appendix B for reference. The SMarch pattern requires a total of 24mN cycles, where "m" is the number of bits in each word.

8. PSEUDO-RANDOM PATTERNS

Pseudo-random memory test patterns involve applying pseudo-random stimulus to some or all of the inputs of the memory [132]. The pseudo-random input stimuli are generated by a linear feedback shift register or LFSR. Another name for a LFSR is a PRPG or pseudo-random pattern generator. When pseudo-random patterns are applied to the inputs of a memory the outputs are observed into an LFSR-like circuit called a multiple input signature register or MISR.

Table 9-11. Example three bit pseudo-random sequence.

Pass	X[0]	X[1]	X[2]
1	1	0	0
2	1	1	0
3	1	1	1
4	0	1	1
5	1	0	1
6	0	1	0
7	0	0	1
8	1	0	0
9	1	1	0
10	1	1	1
11	0	1	1

Pseudo-random patterns are referred to as such since they are random in appearance. In actuality the patterns are quite deterministic, it is just that they appear random. Random patterns would never repeat but pseudo-random patterns repeat the same patterns in the same sequences over and over again. Table 9-11 displays a three bit pseudo-random sequence. Please note that while the patterns appear reasonably random the first and the eighth sequence are identical. Likewise, the second and the ninth are identical, and so on.

It should also be noted that the all zero state is not included. Normal LFSRs skip the all zero state and only with special logic can it be achieved. A five bit pseudo-random sequence is given in appendix B for reference and example. It has the same characteristics as that of the three-bit sequence shown here.

Pseudo-random patterns may be applied to some or all of a memory's inputs. If pseudo-random patterns are applied to the control inputs of a memory, the read and write operations happen pseudo-randomly. If these type patterns are applied to the address inputs of a memory, the addressing order is performed pseudo-randomly. In this case the addresses are proceeded through in one order; it could be referred to as "incrementing." The order cannot (easily) be reversed to decrement through the address space. A different addressing order can easily be achieved but decrementing cannot. Pseudo-random stimuli can also be applied to the memory data inputs.

Some recommend using pseudo-random patterns to identify defects which were not included in the original list of fault models anticipated during memory design [133]. Pseudo-random patterns are very inefficient as very many patterns are required to be applied to activate and detect certain defects. Furthermore, since the pseudo-random sequences are always the same, if pseudo-random stimuli are applied to all of the memory inputs then certain operation sequences will never be performed. For instance, to detect a deceptive destructive read, two reads must occur at each address with each data type before an intervening write is performed. This sequence will simply not occur for every cell location, thereby missing certain very real manufacturing defects. The recommended back-to-back double read will almost never occur and certainly not on every address. Pseudo-random stimuli may be applied to the memory data inputs while normal deterministic sequences are applied to the address and control inputs. This strategy allows varying data patterns to be encountered by the memory while all of the addresses are reached in normal sequence with the read/write operations being performed as desired. Since most memories have their data bits spread across multiple sub-arrays, the utility of pseudo-random data is not significant. Thus, pseudo-random patterns are lengthy and inefficient at finding defects [134]. Therefore pseudo-random patterns are rarely used other than for a gross design debug phase of development.

9. CAM PATTERNS

A content addressable memory has compare and other logic function deep inside its memory array. Some CAMs allow read back of the memory cells while others do not. Some CAMs include masking functions where a single bit within each entry can be removed from consideration in the compare function [135]. Further, ternary CAMs can allow a single bit in a single entry to have its value removed from consideration in the compare function. Some CAMs include the capability to invalidate all entries in the CAM while others allow individual entries to be invalidated. Each one of these factors drives significant differences in CAM test patterns. Because of these differences, all possible permutations of CAM design topology are not considered here. Instead, base CAM test patterns are provided with associated direction included for handling test of the possible CAM topology permutations.

Table 9-12. Basic CAM compare test sequence.

1. The all cells matching state ensures that the NOR output is not stuck-at a "0" and ensure that a match can be detected.
2. With all cells at a "0" state a single "1" is walked across the compare inputs, which puts a singe XOR output in a "1" state and drives the NOR output to a "0" state. This is illustrated in Figure 9-6.
3. With all cells at a "1" state a single "0" is walked across the compare inputs, resulting in similar tests to step number two above.

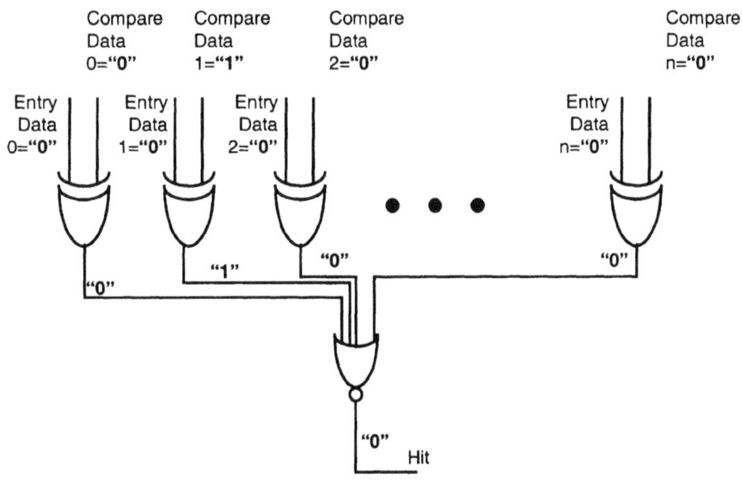

Figure 9-6. CAM compare logic and the resulting test states.

The essential difference between a CAM and a RAM is its compare function. This function is accomplished by the XOR/NOR tree illustrated in Figure 5-2. The memory cells should be tested through standard march patterns using read and write operations. Once the cells are verified as being defect free, the compare circuit can be checked. If the compare function is considered at the gate level of abstraction, as would be typical in logic test, then each XOR needs all four input combinations and the NOR needs to have a single "1" walked across its inputs while the other inputs are maintained at a "0" state [136]. To accomplish this combination each entry needs be examined. The results discussed here are normally applied in parallel to all entries in the CAM. The following operations, shown in Table 9-12, are required to test a basic CAM.

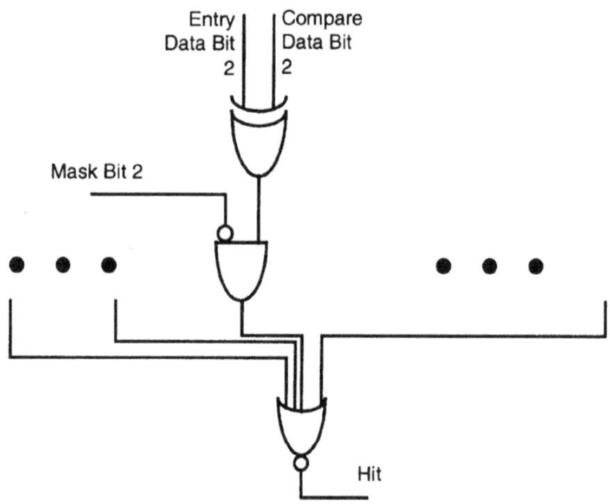

Figure 9-7. CAM compare with masking logic.

Please note that unique data states did not need to be walked across the actual cells. This significantly reduces the test time and the associated test complexity. The pattern shown in Table 9-12 identifies defects in the compare circuitry. Defects in the storage elements of the CAM can be tested through normal marching patterns, assuming that the CAM cells can be read. If they cannot be read then further patterns are required but the CAM compare function can be used to facilitate efficient testing. If there is a mask function in the CAM, the equivalent logic gate diagram would be as shown in Figure 9-7. The AND gates allow any column's cells to be excluded from the match operation. In reality the AND gate is before the XOR, impacting the whole column by forcing both the true and complement compare data lines low [137]. Test is accomplished by having a single bit mismatched, masking that bit, and ensuring that a hit is detected. These along with the many other CAM topology permutations represent more logic-type test issues than memory-specific test issues. The test engineer can determine the topology of their specific CAM design and then apply the needed Boolean combinations to identify stuck-at faults on each of the gate inputs and outputs. The actual transistor operation should, as usual, be considered before finalizing the test patterns. Normally these test patterns can be applied quite efficiently since the CAM entries can be tested in parallel.

10. SOI PATTERNS

Silicon on insulator memories have unique fault models and therefore require some unique patterns. As shown in Figure 4-5, the bodies of the transfer FETs in the six transistor cells electrically float. Since all of the cells along a column can be in one state, all of the transfer FET bodies can float high for one side of the cell, along an entire column. When the bit line on that side of the column must be pulled low, in order to write a single cell, all of the bodies must be discharged along with the bipolar gain, as shown in Figure 4-6. An anemic write head circuit could be insufficient to overcome the current required due to the parasitic bipolar NPN transistors. In addition the write head must be able to drive the bit line capacitance and the larger SOI leakage. Likewise, if during a read the bit line needs to discharge more than a diode drop below Vdd, due to a specific memory architecture, each cell needs to be able to sink sufficient current without disturbing. Because defects can impact the capability of the write head and cell FETs to handle the additional current in SOI technology, a specific pattern is needed [138]. Table 9-13 shows the Walk SOI pattern. The Walk SOI pattern walks a data type up a column with reads, writes, and pauses to accentuate the bipolar currents and thereby identify defects. It should be noted that other patterns are required to capture normal memory defects and that this pattern is targeted to help identify silicon-on-insulator specific defects.

Table 9-13. Description of the Walk SOI pattern.

1	W0 ⇕
2	Pause
3	W1, R1, Pause, R1, W0 ⇑
4	W1 ⇕
5	Pause
6	W0, R0, Pause, R0, W1 ⇑

11. MULTI-PORT MEMORY PATTERNS

Multi-port memories have multiple paths into and out of their memory cells. Since multiple operations can be performed simultaneously, faults which are only activated by multiple operations must have proper patterns generated for them. Patterns which detect faults unique to multi-port memories will exercise the memories with multiple operations in a single cycle. Each port should be exercised with typical memory test patterns to ensure that single port operation can properly be performed. In addition, operations which exercise various ports simultaneously must be performed. Testing for inter-port and intra-port faults must be thorough.

As shown in Figure 8-22 a multi-port memory fault can create problems while writing to one port and reading from another port. If only port A is exercised, then good operation is perceived. As stated earlier, write operations more easily activate defects while read operations more easily detect defects. This difference occurs because write operations require a large swing on a bit line while read operations only require a small swing and the corresponding sensing.

A number of patterns include accessing two ports with different addresses going to each port. Some include a pattern where one port's address sequence is incremented while the other port's address is decremented [139,140]. While this sounds difficult, it can actually be accomplished with a single address counter. The address counter's output is used for the port that is being incremented. The same counter's output is inverted and then used for the port that is being decremented [141]. Other addressing combinations may require more than a single counter, require add/subtract operations, or require one or more stages of address delay. These functions can provide complex addressing pattern sequences which efficiently pursue subtle multi-port memory faults.

An example two-port memory test pattern is given in Table 9-14. The $2PF2_{av}$- pattern addresses two port memory faults [142], where the "2PF" denotes two port memory faults and the second 2 denotes faults that impact two memory cells. The subscript "av" indicates that operations must be performed on the aggressor (a-cell) and victim (v-cell) simultaneously. The "-" denotes an improved version of the pattern with a reduced number of required operations. The term "$\Uparrow$ (c=0..C-1)" denotes incrementing columns while the term "$\Uparrow$ (r=0..R-1)" denotes incrementing rows. A subscript of "r+1" denotes that one of the operations is being performed on the next incremental row. The "&" denotes simultaneous multi-port operation. Some literature utilizes a ":" instead. The "& n" term in the first march element denotes that no operation is performed on the second port. Alternatively the "& n" could have been left off, again denoting no operation on the second port. A comma denotes operations the occur on successive cycles.

The $2PF2_{av}$- pattern illustrates the importance of addressing order in multi-port memories. In the first step of the second march element a write

"1" is performed on the base cell while a read "0" is performed on the next cell up the column, i.e. at the base cell's row plus one. Because of the ways that the cells in a multi-port memory share a column it is important to detect faulty interaction. It can also be seen how the nomenclature employed by this pattern could easily be used to increment rows while decrementing columns, etc.

Table 9-14. Description of the March 2PF2av- pattern.

1	$W0 \ \& \ n \Updownarrow$
2	$\Uparrow(c=0..C-1)[\Uparrow(r=0..R-1)[W1_{r,c} \ \& \ R0_{r+1,c}, R1_{r,c} \ \& \ W1_{r+1,c}, W0_{r,c} \ \& \ R1_{r+1,c}, R0_{r,c} \ \& \ W0_{r+1,c}))]]$
3	$\Uparrow(c=0..C-1)[\Uparrow(r=0..R-1)[W1_{r,c} \ \& \ R0_{r,c+1}, R1_{r,c} \ \& \ W1_{r,c+1}, W0_{r,c} \ \& \ R1_{r,c+1}, R0_{r,c} \ \& \ W0_{r,c+1}))]]$

Another example multi-port memory test pattern is the MMCA or modified March C algorithm [143]. This pattern has the same number of operations for each port as that of the March C algorithm. Table 9-15 displays the pattern. The subscript "i" is the primary port being exercised and brought through the March C pattern. The subscript "∀j" means to perform an operation on all ports "j" simultaneously but this inherently leaves out port "i". A subscript of "-2" means the address the "j" ports are accessing is the "i" port's address minus two. An "x" denotes that the read is a read-don't-care operation. The "&" symbol indicates that multi-port operations on either side of the "&" are to be performed in the same cycle.

Table 9-15. Description of the MMCA pattern.

1	$W_i0 \Updownarrow$
2	$R_i0 \ \& \ R_{\forall j-2}x, W_i1 \ \& \ R_{\forall j+2}x \Uparrow$
3	$R_i1 \ \& \ R_{\forall j-2}x, W_i0 \ \& \ R_{\forall j+2}x \Uparrow$
4	$R_i0 \ \& \ R_{\forall j-2}x, W_i1 \ \& \ R_{\forall j+2}x \Downarrow$
5	$R_i1 \ \& \ R_{\forall j-2}x, W_i0 \ \& \ R_{\forall j+2}x \Downarrow$
6	$R_i0 \Updownarrow$

The second march element has a read "0" being performed on port i while at the same time performing a read-without-compare on all of the other ports. In the second cycle of this march element a write "1" to port i is being performed while again reading all of the other ports. This type pattern detects the normal single-port faults along with a number of multi-port memory interaction faults.

12. SUMMARY

Numerous patterns have been described in this chapter and numerous more have been developed over the years. The key is utilizing the best patterns to detect memory defects. Since it is essential that faulty memories be identified to prevent them from being shipped to the customer, thorough testing must be employed. The wrong patterns will allow bad chips to pass test.

The patterns described in this chapter do not cover all possible memory topologies. New memory configurations are generated each year as can be seen at any circuit design conference. The key is understanding fault modeling and the pattern sets described here. With a thorough examination of the specific transistor configurations utilized in the memory of concern, the appropriate fault models can be selected and the proper patterns generated. Patterns should not be considered a menu to choose from but rather a starting point for developing the correct pattern for a given memory.

Appendix B includes further patterns that can be examined for reference. Some are theoretically interesting while others can provide very helpful insight. Tables 9-16 and 9-17 describe the key factors in memory test patterns and the primary nomenclature, respectively.

Table 9-16. Summary of factors in memory test patterns.

n	Number of bits
N	Number of address locations
m	Number of bits in a word
R	Number of rows
C	Number of columns

Table 9-17. Nomenclature for test patterns.

R0	Read "0"
R1	Read "1"
Rx	Read don't compare
W0	Write "0"
W1	Write "1"
$\Uparrow$	Increment through addresses
$\Downarrow$	Decrement through addresses
$\Updownarrow$	Addressing order is non-consequential
,	Perform operations on successive cycles
&	Perform simultaneous multi-port operation
R_i	Read port i
$R_{\forall j}$	Read all ports j, where j does not include i
R_{-2}	Read at the current address minus two

Chapter 10

BIST Concepts

Memory Self Test

"His mind was crowded with memories; memories of the knowledge that had come to them when they closed in on the struggling pig...." – from *Lord of the Flies*

Memory built-in self-test, or as some refer to it array built-in self-test (ABIST), is an amazing piece of logic. A very complex embedded memory, without any direct connection to the outside world, is tested easily, thoroughly, and efficiently. The memory quality, through the use of a small on-chip tester, is improved and redundancy implementations are calculated. Test issues are naturally pulled into the design team's thoughts since a BIST needs to be designed onto the chip. Test, in this manner, is developed concurrently with the rest of the design. BIST allows shorter test time since cycle-after-cycle testing is provided by the BIST, which is impossible by any external means. Thus, memory built-in self-test is key and will only become more so as more embedded memories are included on each chip.

Proper self test of memories, however, requires an understanding of both memory design and memory test. The first part of the book described various memory design techniques to aid the reader in understanding individual memory topologies and how they can fail. The second part of the book dealt with memory fault models and test patterns, all being fundamental to memory testing. The understanding of both design and test must be clear to develop a good memory self-test strategy. Often, neither a complete understanding of memory design nor of memory test is involved and instead an arbitrary memory test pattern is implemented in logic. The results are anything but stellar.

Memory built-in self-test is a highly optimized test approach [144]. To accomplish this, a good BIST engine must be provided for the memory and there are a number of forms this can take. Regardless of form, it is necessary that the performance and area of the memory not be adversely impacted [145,146]. The memory boundary is crucial, not only to provide functional and BIST inputs but also to avoid memory performance degradation. This chapter starts with a discussion of the interface boundary and then provides an introduction to BIST techniques and issues.

1. THE MEMORY BOUNDARY

The memory must be thoroughly tested in a microscopic area and with minuscule performance impact. Much of stand-alone memory testing surrounds the concept of testing the chip I/O. Since BIST normally involves an embedded memory there are no chip I/O. Therefore, the I/O in terms of voltage potential levels, drive, and impedance can all be ignored.

The I/O to an embedded memory is really a logic interface which utilizes full swing values. Thus, the interface testing has been vastly simplified, making the concept of BIST tractable.

The boundary through which the BIST accesses the memory should be clean. In other words there should be a minimum of logic inside the boundary, i.e. in between the boundary and the memory array itself. In this manner the BIST engine can apply the needed patterns without having to concern itself with conditioning preceding logic. Similarly, the output of the memory should be with a minimum of logic in between the memory array and the output compare or compression.

In testing the embedded memory, it is important to remember that the chip also sees logic testing, for the logic portions of the chip. The logic test and memory test should at a minimum meet and in all likelihood overlap to ensure that all parts of the chip are tested. The logic leading up to the memory and the logic leading away from the memory all need to be fully tested to ensure adequate quality. This normally cannot be done by memory BIST. Figure 10-1 shows a BIST engine and a memory with representative signals in between them. The data input, address, and controls are provided to the input of the memory. The output from the memory is compared and a pass/fail indication is sent back to the BIST engine. There are a number of permutations that are possible on the BIST to memory interface. Nonetheless this figure illustrates the essence of a memory BIST and its associated memory. Please note that the figure is not to scale. The memory BIST is normally quite small in comparison to the memory it is testing.

At the input of the memory the interface can be as simple as a multiplexer or a scannable latch. A latch is often employed to aid in controlling set up and hold times to the memory and can greatly ease testing.

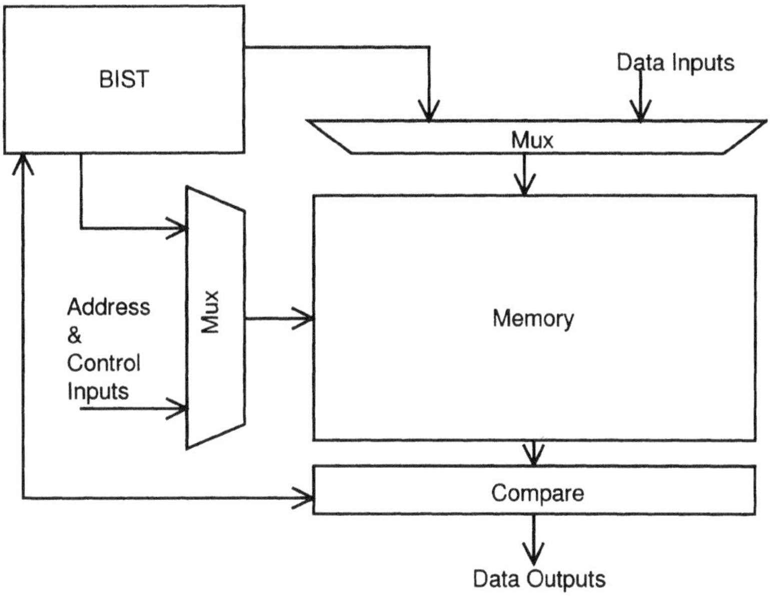

Figure 10-1. Memory to memory BIST interaction.

If a multiplexer is utilized, it has one input which comes from the functional path. The other input comes from the BIST engine. A select signal for the multiplexer selects the BIST path, when the memory is under test. A slightly more complex arrangement is one where the output of the multiplexer is routed to an observation latch. In this manner, the logic test boundary continues beyond the multiplexer output into the observation latch. The multiplexer can, in this way, be fully tested. During logic test, signals are sent from the BIST engine and captured into the latch. Furthermore, during logic test signals are sent along the functional path and captured into the multiplexer. Thus the I/O of the multiplexer are fully tested during logic test. During memory BIST the path from the multiplexer to the memory is exercised. In this manner, all parts of the multiplexer and paths into the memory are tested by a combination of memory BIST and logic test. This multiplexer arrangement can be utilized on all data, address, and control inputs to the memory. The output of the multiplexer can be observed, either directly or after some combinational logic, illustrated by the bubble in Figure 10-2. The input from BIST can go directly to the multiplexer or it can be used as shown here. In this example the signal from BIST is latched and can be fanned out to other memory inputs. This configuration can reduce the number of BIST to memory signals and improve the speed at which the memory/BIST combination can run.

Synchronous memories, which almost all embedded memories are, require a clock input. The clock that exercises the memory during BIST can be the same clock as that used during functional operation; this method is generally preferred. Alternatively, there can be a second clock supplied to the memory during BIST test which goes through a multiplexer-type arrangement. Obviously, the key thing is to get a clock to the memory for test. The clocking strategy is primarily driven by the logic test strategy and performance related issues.

The memory output requires a compare or compression circuit. The selection of one versus the other depends on whether deterministic or pseudo-random patterns are being applied. It is also driven by the presence of redundancy and diagnostics issues, which are covered later.

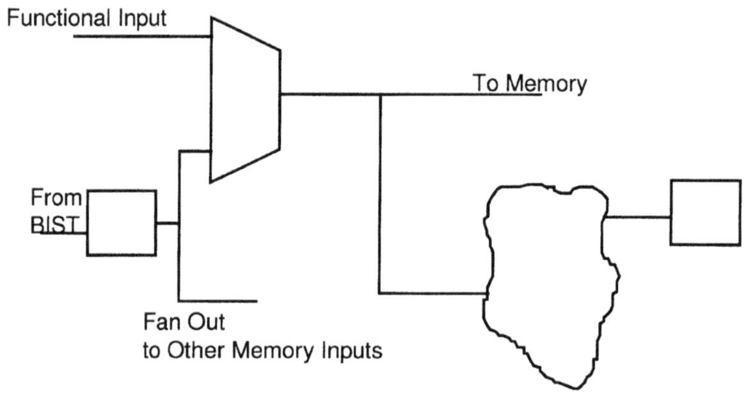

Figure 10-2. Memory input interface.

2. MANUFACTURING TEST AND BEYOND

Memory built-in self-test provides a means for testing a memory. BIST testing can be accomplished either in manufacturing or in system [147]. In manufacturing, BIST can be utilized at wafer test, module test, card test, and initial system test. System test, using BIST, can be performed at each system power or it can be performed occasionally while the system is running. If BIST is utilized while the system is running, the data contents from the memory must be temporarily placed in some other location while the BIST is exercised. After test is complete the data must be restored to the memory. This type of testing operation can be referred to as transparent BIST [148,149].

BIST is highly optimized for embedded memories but there are times when BIST makes sense for stand-alone memories as well. Normally, a stand-alone memory is tested on a memory tester at time of manufacturing. BIST normally is used in conjunction with a logic tester. If there is a memory tester available it doesn't make sense to use BIST on a stand-alone memory, since the I/O are available. The large ATE memory tester is far more powerful than a BIST and therefore should be utilized for stand-alone memories but not for embedded ones.

When a stand-alone memory is used in a multi-chip module (MCM) the tester for the MCM is a logic tester. Having BIST on the memory chip enables test of the memory in this logic environment, which would otherwise be impossible or at least vastly inefficient and of limited quality. Thus BIST can be included on the stand-alone memory to facilitate its test during MCM manufacturing [150]. Further, that MCM memory can be tested in the field with the very high quality test provided by BIST. The quality of a BIST test is vastly superior to any system power-on test that would be applied to the memory by the operating system and thus provides significant advantage. This advantage can be utilized with soft redundancy and will be discussed more in chapter 13.

The other place that BIST, or a form of it, can be utilized with stand-alone memories is when a chip needs to thoroughly test memories that are attached to it through a bus. Perhaps built-in self-test should not be the term utilized but this has caught on in the industry. Some refer to this as external BIST or some similar name. Being both "external" and "built-in" sound contradictory and indeed are. Nonetheless, a test engine which is very similar to a normal BIST engine is utilized to generate patterns to test these off-chip memories. One of the special tests applied for these external BIST tests is signal continuity and signal shorting, ensuring that the large data bus and address bus going between the memories is indeed intact.

3. ATE AND BIST

Frequently there are discussions, more often debates, on the use of automated test equipment (ATE) versus the use of BIST. In the case of embedded memories, BIST is the only practical solution. Large external testers cannot provide the needed test stimulus to enable high speed nor high quality tests [151]. ATE is still critical, though, since it is used to initialize, clock, and read out the BIST. It is also through ATE that the BIST itself is tested or at least facilitated to be tested. The BIST is made up of logic which must be tested, either by automatic test pattern generation (ATPG) means or by logic BIST. If the memory BIST logic has a manufacturing defect and is not tested, a failing memory can be identified as passing. Clearly, the BIST must be tested thoroughly. Lastly, ATE is key to enabling diagnostics with BIST. A cycle-by-cycle pass/fail flag can be used to generate an address fail map. This can only work if ATE is functioning to capture the result as it is generated cycle-by-cycle with the BIST. BIST has great capabilities but they can only be enabled, at least initially, via ATE.

4. AT-SPEED TESTING

Testing of certain faults in memories requires high performance at-speed testing. Ideally a BIST should be able to run much faster than the memory it is intended to test. In this manner, especially during initial design characterization, the memory can be pushed to its limit without breaking the capability of the BIST engine. Depending on memory topology, many faults can be caught without at-speed, often referred to as DC, testing. For memories with self-timed latched sense amplifiers, all of the circuits except for the pre-charge FETs are tested through slower testing. Still, the only way to ensure that the bit lines pre-charged correctly is to run the memory with full speed, or AC, test. In addition, more noise is generated during full speed clocking and noise can defectively be injected into the sense amplifiers. This noise problem can only be found with at-speed testing or worse it can cause a system fail.

Some memories do not use latched sense amplifiers, in which case slower tests provide very little AC coverage. For DRAMS, a memory cell is read and then a write-back is performed to restore the cell to its original value. In this case cycle time is a very key test parameter and at-speed testing is required. Running a DRAM slowly allows extra time for the write-back to occur, which results in a poor AC test. Thus an at-speed test is very desirable and utterly necessary for certain memory types.

5. DETERMINISTIC BIST

A built-in self-test engine, no matter what kind, generates deterministic patterns. Deterministic means that the patterns generated follow specific pre-determined values. These pattern values are defined by the logic of the BIST design. The opposite of deterministic are patterns that are random. No BIST patterns are truly random. Even those patterns which are pseudo-random are not random but follow specific sequence defined by the logic. Thus pseudo-random patterns are deterministic, but more is discussed on this in the next section.

Deterministic also generally means that the BIST generates algorithmic patterns along the lines of those described in chapter nine and appendix B. For example the March C- pattern, as described in Table 9-5 accesses each cell 14 times. The operations performed in the second element are a read "0" followed by a write "1" on each addresses. The addresses are proceeded through sequentially. Thus this specific, regular pattern is deterministic. The BIST generates these patterns and this is the norm for most built-in self-test engines. The next two chapters will cover more on the BIST engines that generate these patterns.

6. PSEUDO-RANDOM BIST

A pseudo-random pattern is very helpful for testing random logic [152]. A memory, however, is a regular structure and needs the application of regular patterns. In the early days of memory BIST it was not unusual to see a pseudo-random patterns applied [153] but virtually no one uses these today in a manufacturing environment. Using pseudo-random patterns does not provide double back-to-back reads, which are effective at finding read disturb defects. They also will not provide a physical checkerboard pattern which helps find coupling types of defects. Thus, many memory defects are missed by pseudo-random testing. Pseudo-random patterns are still utilized on an occasional basis in characterization of a design. The pseudo-random test applies patterns which otherwise would not be considered.

The use of pseudo-random patterns during manufacturing test is a hopeful attempt to catch some defective operation that would otherwise be overlooked. Fortuitous testing should not be counted on in a manufacturing environment and thus pseudo-random patterns should not be emphasized. The key in memory testing is understanding the circuit arrangement, selecting the proper fault models, and then testing for those specific faults. If this is done well then the memory tests will be sufficient and the resulting chips will be of high quality. For these reasons pseudo-random testing doesn't really have a place in good BIST practice for most memories.

Key point: Pseudo-random patterns do not provide the needed memory test quality.

From a historical point of view, it is helpful to understand pseudo-random BIST. Since logic BIST employs pseudo-random techniques many people are somewhat familiar with the concepts and those which can be safely applied to memories should be articulated.

A pseudo-random pattern is generated by a linear feedback shift register or LFSR. Another name for a LFSR is a pseudo-random pattern generator or PRPG. A LFSR employs a series of latches and XOR gates to form the logic. The latches and XOR gates are constructed based on a primitive polynomial that ensures all of the 2^n-1 states are exercised, where "n" is the number of latches that forms the LFSR. The primitive polynomial provides a maximum length sequence. The only state which is not included is the all zeros state and special logic must be included in the LFSR if the all zeros state is required. Primitive polynomials can be looked up in a number of references [154]. The primitive polynomial for a 3 bit LFSR is $x^3 + x + 1$. That means that the taps for the XORs are on the X3 and X1 bits. Much literature starts with a X0 bit rather than an X1 bit, where the X0 bit represents x^0. It is sometimes easier to implement the resulting logic when starting to count with the X1 bit. The x^3 and x terms corresponds to an XOR tap on X3 and X1 bits respectively, as shown in Figure 10-3. The "1" term is in every primitive polynomial and drives no added XOR taps in the logic. It should be known that following proper mathematics yields latches numbered X^0, X^1, and X^2. This LFSR generates the pattern that was shown in Table 9-11.

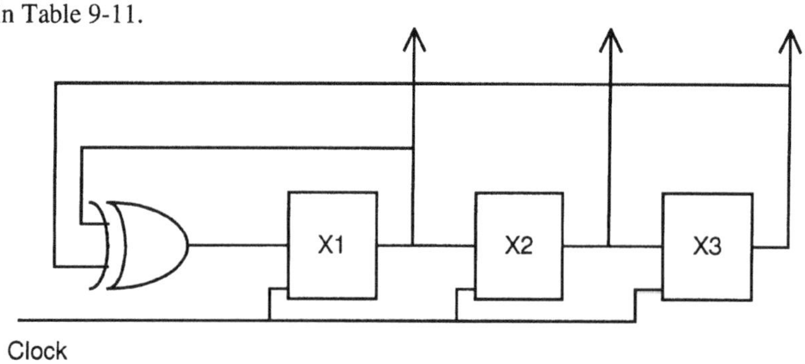

Figure 10-3. Example three-bit LFSR.

Another example has a nine-bit primitive polynomial of $x^9 + x^4 + 1$. From this the XOR taps are from the X9 and X4 bits. The resulting LFSR design is shown in Figure 10-4. The first 16 cycles of this LFSR are shown in Table 10-1. A later group of sixteen cycles, including the 512th cycle, which is a repeat of the first cycle's pattern, is shown in Table 10-2.

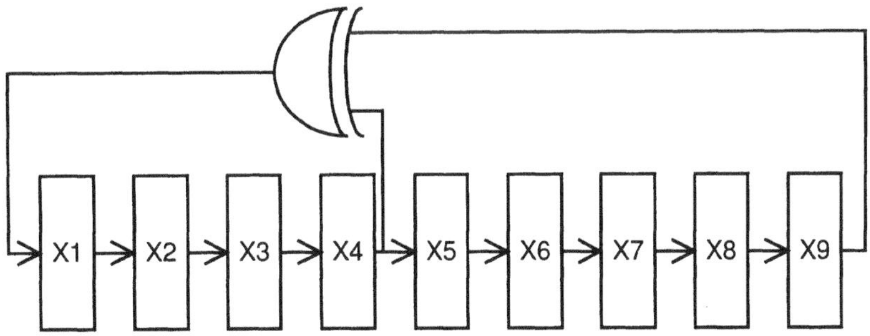

Figure 10-4. Example nine-bit LFSR.

Table 10-1. First 16 cycles of the nine-bit LFSR.

Pass	X1	X2	X3	X4	X5	X6	X7	X8	X9
1	1	0	0	0	0	0	0	0	0
2	0	1	0	0	0	0	0	0	0
3	0	0	1	0	0	0	0	0	0
4	0	0	0	1	0	0	0	0	0
5	1	0	0	0	1	0	0	0	0
6	0	1	0	0	0	1	0	0	0
7	0	0	1	0	0	0	1	0	0
8	0	0	0	1	0	0	0	1	0
9	1	0	0	0	1	0	0	0	1
10	1	1	0	0	0	1	0	0	0
11	0	1	1	0	0	0	1	0	0
12	0	0	1	1	0	0	0	1	0
13	1	0	0	1	1	0	0	0	1
14	0	1	0	0	1	1	0	0	0
15	0	0	1	0	0	1	1	0	0
16	0	0	0	1	0	0	1	1	0

Table 10-2. Sixteen later cycles including the start of the repeat.

Pass	X1	X2	X3	X4	X5	X6	X7	X8	X9
499	0	1	0	0	0	1	1	0	0
500	0	0	1	0	0	0	1	1	0
501	0	0	0	1	0	0	0	1	1
502	0	0	0	0	1	0	0	0	1
503	1	0	0	0	0	1	0	0	0
504	0	1	0	0	0	0	1	0	0
505	0	0	1	0	0	0	0	1	0
506	0	0	0	1	0	0	0	0	1
507	0	0	0	0	1	0	0	0	0
508	0	0	0	0	0	1	0	0	0
509	0	0	0	0	0	0	1	0	0
510	0	0	0	0	0	0	0	1	0
511	0	0	0	0	0	0	0	0	1
512	1	0	0	0	0	0	0	0	0
513	0	1	0	0	0	0	0	0	0
514	0	0	1	0	0	0	0	0	0

There are LFSRs that have distributed feedback, where the XORs are at the inputs of multiple latches. The alternative is shown in these examples where the XOR is only at the input of the first latch. Each can be used but one may be easier to implement for a specific application.

Once the needed LFSR has been generated, its outputs can be connected to the inputs of the memory. Lets look at a simple example of testing a 64-bit memory and using the previous LFSR to provide test stimulus. In the case of the nine-bit LFSR, the first output could be the read/write control. The second bit could be the data input bit. The third through eighth bits could be the address inputs. The ninth LFSR bit could be disconnected. By connecting only eight out of the nine bits, it can be assured that the all-zero state is achieved for those eight bits. With a normal LFSR, the all zero state cannot be achieved but n-1 zeros are always achieved no matter which LFSR bit is not connected. The pseudo-random patterns applied to the inputs of the memory are shown in Table 10-3. With the connections described the first cycle is a read of the 0th address. This assumes that when the read/write control is high, a read is performed. When it is low a write is performed. The next cycle is a write of a "1" to the 0th address and this process continues until the test is complete. On the tenth cycle, a read of the 4th address occurs. The data in the 4th address is a "1" since an earlier cycle, i.e. the sixth cycle in this case, wrote a "1" to that address.

If one follows this example the sequence of patterns applied to the memory repeats every 511 cycles. It should also be noted that there is always a read of address zero followed by a write of a "1" to address zero, and so on. This means, as mentioned earlier, that there will not be back-to-back double reads performed on each address for instance. Other test sequences are not performed either and thus various fault models are not activated during a pseudo-random test [155]. If the test were truly random then with sufficient time and cycles, all of the possible sequences would be performed and the needed tests would be generated. Since the tests are pseudo-random instead of random, the same sequences are repeated over and over again.

The example above utilizes a single LFSR to provide test stimulus to the memory. It is possible to have more LFSRs, where one is used for data and control while another is used for address generation. The LFSRs would be of different lengths and therefore would have different repeat factors. With this configuration the stimulus appears more random and therefore with sufficient time and cycles more of the possible faults would be detected. Further it is possible to use a LFSR on some of the inputs while generating other inputs in a standard deterministic fashion. A typical example would be to connect the address inputs to a LFSR and then generate the control and data deterministically. It should be noted that there is still no method included here to increment through the addresses in one fashion and then decrement in the reverse fashion without a very complex LFSR structure.

Table 10-3. Pseudo-random testing example of a 64-bit memory.

Pass	R/W	Data	A0	A1	A2	A3	A4	A5	X9
1	1	0	0	0	0	0	0	0	0
2	0	1	0	0	0	0	0	0	0
3	0	0	1	0	0	0	0	0	0
4	0	0	0	1	0	0	0	0	0
5	1	0	0	0	1	0	0	0	0
6	0	1	0	0	0	1	0	0	0
7	0	0	1	0	0	0	1	0	0
8	0	0	0	1	0	0	0	1	0
9	1	0	0	0	1	0	0	0	1
10	1	1	0	0	0	1	0	0	0
11	0	1	1	0	0	0	1	0	0
12	0	0	1	1	0	0	0	1	0
13	1	0	0	1	1	0	0	0	1
14	0	1	0	0	1	1	0	0	0
15	0	0	1	0	0	1	1	0	0
16	0	0	0	1	0	0	1	1	0

When pseudo-random testing is performed on a memory, the outputs cannot be compared with expected data. Since the data was written into the memory in a pseudo-random fashion, the output is not very predictable. Another pseudo-random technique is utilized for compressing the output results and determining whether the resulting test passed or failed. A multiple input signature register or MISR can take the data output values read from the memory during the entire test sequence and generate a string of bits that indicate a pass/fail condition. A MISR is very similar to a LFSR, with only the addition of an XOR gate for each data output bit being read. Thus if a memory had nine data outputs, the nine-bit LFSR that was described earlier in Figure 10-4 could be made into a MISR, as shown in Figure 10-5. If only a single data output exists then a single input signature register or SISR would be utilized. Again, it is similar to a LFSR but in this case a single XOR is provided for observing the output into the compressor. When a MISR or SISR is utilized a passing signature must be generated. A simulation is performed which loads the results of the reads into the register. A new pseudo-random number is generated for each cycle. At the end of the simulation the anticipated resulting signature is obtained. If that result is found at the end of a test then the memory is said to be good. When using a MISR, the memory must first be initialized to known values to prevent reading an "X" into the MISR. Unknown values or Xs corrupt the MISR signature.

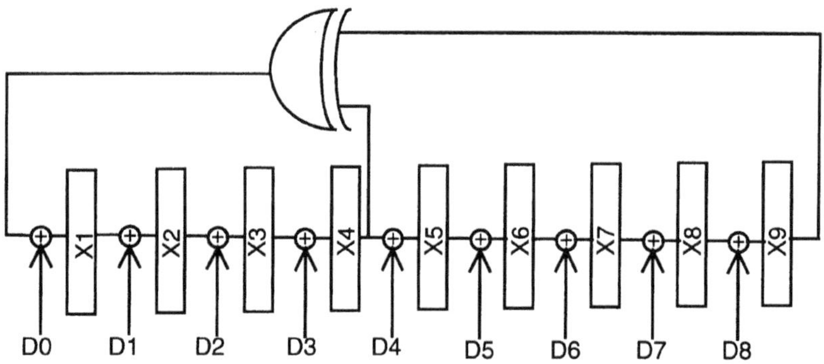

Figure 10-5. A nine-bit MISR schematic.

It is important that the MISR or SISR be of sufficient length to prevent aliasing [156]. An alias can occur when a failing signature is permuted back into a passing signature. With a short MISR or SISR there are very few pseudo-random numbers. It would be easy for a failing result, therefore, to modify one of those numbers back into the anticipated value. Another way that aliasing can occur is to have two errors. The first causes a bad signature and the second causes the bad signature to be modified back into a good signature. Table 10-4 shows a good memory simulation for the state of a nine-bit MISR. The starting signature is 100000110, which has been chosen arbitrarily. Let's assume that the MISR is working from left to right and that the memory is doing a read of zeros on each cycle. The faulty memory result that aliases is shown in Table 10-5. In this case the third and fourth cycles each have an error and those errors are one bit apart. When the D5 bit is read on the third cycle, the corresponding MISR bit flips to an erroneous value. On the following cycle the 6th bit is in error. The MISR's D6 bit input XORs the D5 bit's output with the memory's 6th bit output, re-inverting the failing signature input to the passing signature. Please note the highlighted values for D5 and D6 in the 3rd and 4th pass, respectively. Note also that the subsequent signatures match those from a good memory simulation. Obviously, this kind of erroneous operation would be very rare but it is important, nonetheless, to understand. If no further failures are encountered, the MISR falsely indicates that the memory is defect free.

Table 10-4. Passing memory MISR signature.

Pass	D0	D1	D2	D3	D4	D5	D6	D7	D8
1	0	1	0	0	0	0	0	1	1
2	1	0	1	0	0	0	0	0	1
3	1	1	0	1	0	0	0	0	0
4	1	1	1	0	1	0	0	0	0
5	0	1	1	1	0	1	0	0	0
6	1	0	1	1	1	0	1	0	0
7	1	1	0	1	1	1	0	1	0
8	1	1	1	0	1	1	1	0	1
9	1	1	1	1	0	1	1	1	0
10	1	1	1	1	1	0	1	1	1

Table 10-5. Aliasing of failing signature.

Pass	D0	D1	D2	D3	D4	D5	D6	D7	D8
1	0	1	0	0	0	0	0	1	1
2	1	0	1	0	0	0	0	0	1
3	1	1	0	1	0	1	0	0	0
4	1	1	1	0	1	0	0	0	0
5	0	1	1	1	0	1	0	0	0
6	1	0	1	1	1	0	1	0	0
7	1	1	0	1	1	1	0	1	0
8	1	1	1	0	1	1	1	0	1
9	1	1	1	1	0	1	1	1	0
10	1	1	1	1	1	0	1	1	1

A MISR can be utilized to observe memory outputs when normal deterministic test patterns are applied to memory inputs. Furthermore, a MISR is normally utilized when BIST testing a ROM, in conjunction with an up/down counter. The MISR captures the results of all the test cycles and accumulates a signature. If the signature, at the end of test, matches the simulated good-memory result, the memory passes test. A MISR cannot provide a cycle-by-cycle pass/fail result but only provides a pass/fail determination at the end of test. It is possible to determine the pass/fail signatures at intermediate cycles and therefore determine when the memory failed. Nonetheless it is not possible to utilize a MISR with BIST if a memory has redundancy since the BIST must know in which cycle the memory fails.

Pseudo-random memory test techniques, thus, can be useful in memory result compression but not in pattern generation. The logic is relatively simple but the quality of the test patterns is poor and test time is excessive.

7. CONCLUSIONS

A BIST engine generates patterns to provide the test of a memory. A good BIST engine provides thorough test. "Any old" BIST is not sufficient, though, to provide this thorough test. Instead the BIST must test according to an understanding of the memory design, properly developed fault models, and optimal test patterns.

The BIST can be designed as a finite state machine or it can be designed as a micro-code BIST. These two types will be covered in the next chapters. Following this will be a discussion of how BIST handles redundancy and of other design-for-test and BIST techniques which help in the test of memories.

Chapter 11

State Machine BIST

Memory Self Test

"That is why you fail." – Yoda in *The Empire Strikes Back*

A state machine BIST exercises a memory with a pre-determined set of patterns. This type of BIST goes by the name state machine, finite state machine, or FSM [157]. These names are often used interchangeably.

A state machine BIST can generate a single simple pattern or a complex suite of patterns [158]. Most BISTs in industry only generate a single pattern like one of those shown in chapter nine [159]. A sweep through the address space writing to the memory array, followed by a series of reads and writes, and finally an address sweep of reads is performed. This sequence is controlled by a group of operation loops performed by the state machine BIST.

While a single pattern is typical, a better memory test solution is to generate a suite patterns with a state machine BIST [160,161]. The suite of patterns drives more complexity in the state machine BIST but catches the faults that real manufacturing defects can cause. A very convoluted set of patterns can be generated by a state machine BIST, making it a powerful tool for catching subtle defects.

A state machine BIST, as the name implies, can exist in a number of states. There are a group of latches, from very few to several hundred, which define the possible states of this BIST [162]. Between the latches is combinatorial logic which determines the next state for each of the latches and thereby the resulting state of the state machine [163].

1. COUNTERS AND BIST

A counter is a key component of any state machine BIST [164] and is found in an address generator, as well as other components. Since most of the patterns described in chapter nine have both increment and decrement portions, an address counter needs to be an up/down counter. The other portions of the state machine BIST engine are counters as well. The difference is the combinational logic that controls when to increment and the state into which to increment.

The state machine for the BIST can be simple or complex. A simple test pattern requires only a simple BIST engine. A BIST engine which provides a suite of test patterns obviously is far more complex. The memory design must be understood and then the appropriate fault models determined [165]. The needed pattern set must be defined, which enables detection of the possible memory fault models. Then the BIST engine can be defined. A thorough BIST engine must inherently be complex if it is going to prevent defective memories from being going to the customer.

2. A SIMPLE COUNTER

Given that the basic component of a BIST state machine is a counter, lets examine a simple one in some detail. An address counter need only increment, if a zero-one pattern as defined in Table 11-1 is desired. For an eight-address memory a three-bit counter can suffice. The needed states are given in Table 11-1 and are represented in the state diagram shown in Figure 11-1.

Table 11-1. State sequence for three-bit address counter.

Adress	Bit(2)	Bit(1)	Bit(0)
0	0	0	0
1	0	0	1
2	0	1	0
3	0	1	1
4	1	0	0
5	1	0	1
6	1	1	0
7	1	1	1

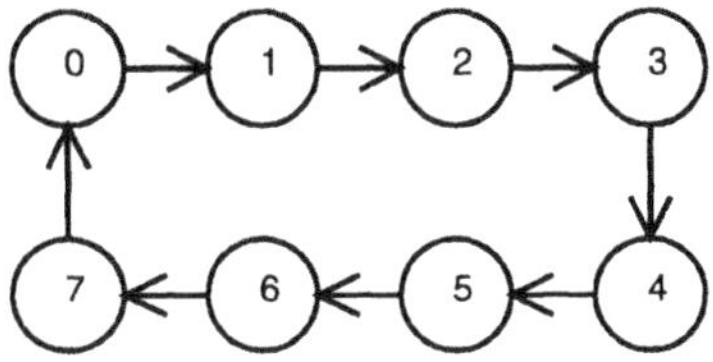

Figure 11-1. State diagram for three bit counter.

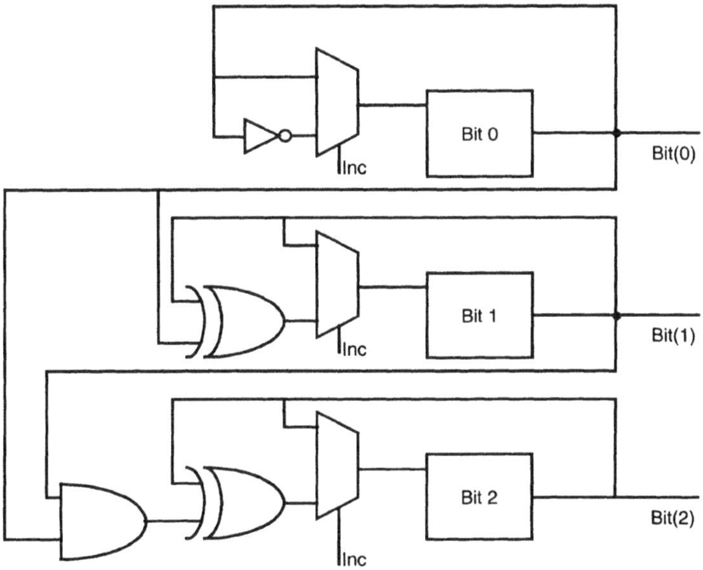

Figure 11-2. Three-bit address counter gate schematic.

For the zero-one pattern, the BIST needs to sweep through the memory address space four times. Clocking can continue and the address counter be allowed to roll over and continue. A simple gate level schematic of a three-bit ripple counter is given in Figure 11-2. The input enable ("inc" for increment) stays high until the address counter completes four full sweeps, after which time it is brought low. When the enable signal goes low it locks up the state machine, stopping the BIST from continuing. It is possible to simply let the BIST engine continue as long as it is clocked but this is generally considered a sloppy design style. A high level design language like Verilog or VHDL can be used to describe a piece of logic like this counter. VHDL defining the three-bit counter operation is given in Table 11-2 [166]. The full counter model is given in appendix C.

Table 11-2. VHDL for a three-bit counter.

```
with inc select
        u_bit(0) <= not Bit(0) when '1',
                    Bit(0) when others;

with inc select
        u_bit(1) <= (Bit(0) XOR Bit(1)) when '1',
                    Bit(1) when others;

with inc select
        u_bit(2) <= ((Bit(0) AND Bit(1)) XOR Bit(2)) when '1',
                    Bit(2) when others;
```

3. READ/WRITE GENERATION

While a counter clearly can generate the needed memory address inputs during BIST, similar logic arrangements can handle the other BIST portions as well. The first two elements of the March C- pattern are given in Table 11-3. In order to implement these, a write-enable signal needs to be activated for the first march element. Alternating read-enable and write-enable signals need to be activated for the second sweep. A state diagram describing this operation is given in Figure 11-3. Note that when the maximum address (Max_Addr) is reached, the next element's operation is started. Once the "End" state is reached a signal goes to the remainder of the BIST to indicate completion or to initiate the next pattern sequence.

Table 11-3. First two march element of the March C- pattern.

1	W0 ⇕
2	R0, W1 ⇑

A short memory BIST simulation is given in Figure 11-4. The example memory has four addresses. The read and write-enable signals, defined to be unique signals for this memory, are active high. Example Verilog, which performs the read and write-enable generation, as well as the data generation is provided in Table 11-4 [167]. (The complete Verilog code for this block is given in appendix C.) It is assumed that the data input signal is also utilized as the expect data signal for comparison at the memory outputs. The gate level logic that generates the read-enable, write-enable, and data signals is given in Figure 11-5.

The data generation can easily be enhanced to provide alternating data based on address. If a data pattern is desired which places opposite data on each address increment, then the least significant bit (LSB) from the address counter can be supplied to the data generator. This type arrangement can be quite helpful in generating a checkerboard pattern. Expanding this concept, the LSB for the row and the LSB for the column can be supplied, all getting XORed with the base data pattern type. Other modifications to the data generator can be utilized to provide even more complex data patterns.

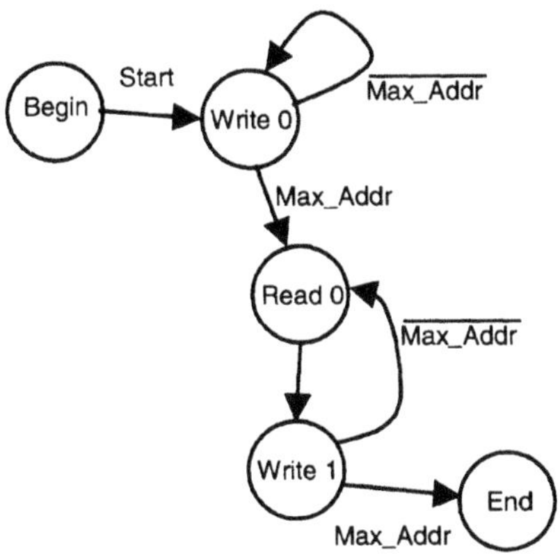

Figure 11-3. State diagram for simplified data and read/write controller.

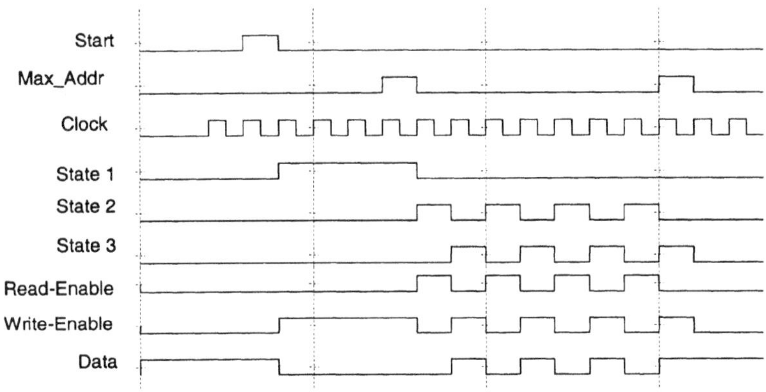

Figure 11-4. Waveforms from simple data and read/write controller.

Table 11-4. Verilog of simple data and read/write controller.

```
always @ (start or state1 or max_addr)
    if ( start ) next_state1 = 1 ;
    else if ( max_addr ) next_state1 = 0 ;
    else next_state1 = state1;

always @ (state1 or state3 or max_addr)
    if ( max_addr & state1 ) next_state2 = 1 ;
    else if (!max_addr & state3 ) next_state2 = 1 ;
    else next_state2 = 0 ;

always @ (state2 or max_addr)
    if ( state2 ) next_state3 = 1 ;
    else next_state3 = 0 ;

assign write = state1  | state3;
assign read  = state2;
assign data  = !(state1 | state2);
```

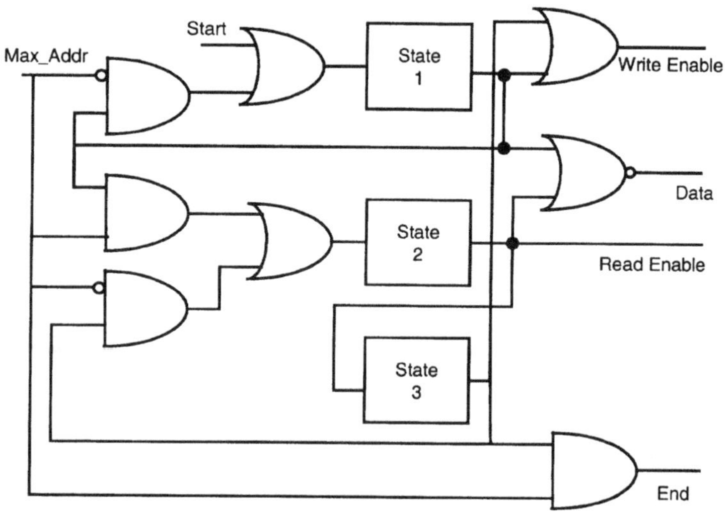

Figure 11-5. Simplified data and read/write controller.

4. THE BIST PORTIONS

A BIST engine can be seen as a combination of blocks. Each block interacts with the others to enable the needed BIST patterns. The primary blocks are the address generator, the data generator, and the read/write control generator, as shown in Figure 11-6. At the start of BIST, the state machine is initialized by scanning, through a test access port, or by a reset control. Once initialization is complete, a clock causes the BIST engine to increment and proceed from state to state.

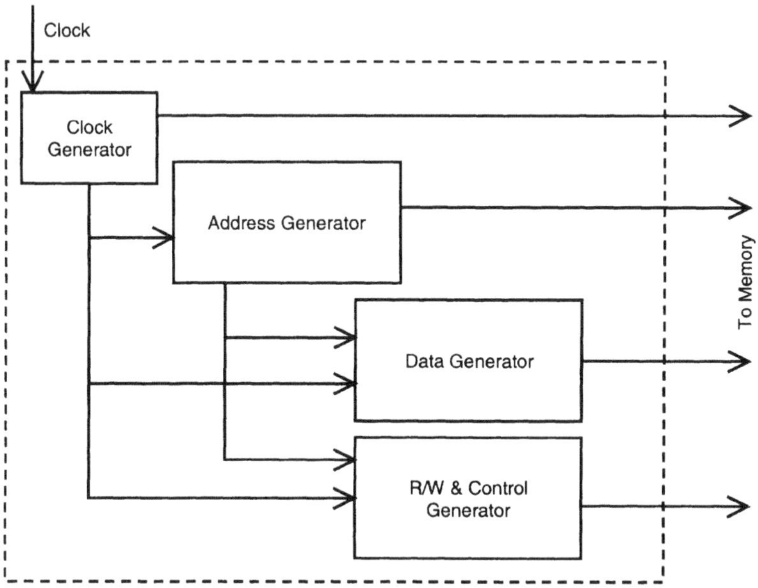

Figure 11-6. Simple state machine BIST block diagram.

A slightly more detailed diagram of a state machine BIST is given in Figure 11-7 with the appropriate BIST to memory interactions. This diagram shows a pattern controller, which defines overall control and determines when the BIST engine proceeds from one pattern to the next [168]. The pattern controller interacts with the read/write controller to allow the correct series of reads and writes for each unique pattern. It also interacts with the data generator and address counter to provide the correct data and address stimuli to the memory.

The address counter interacts with the address limiter and address comparator, identifying the proper start and stop address points. The address comparator interacts with the pattern controller to help identify when one pattern is finished. The read/write controller determines how many cycles a given address is maintained before the address counter is incremented. The address counter indicates if an even or odd address is being exercised, so that the appropriate data can be generated. This counter needs to be able to increment and decrement. Further, it should be able to increment/decrement either row or column addresses most frequently. (Appendix C includes an example address counter with these capabilities in behavioral Verilog.)

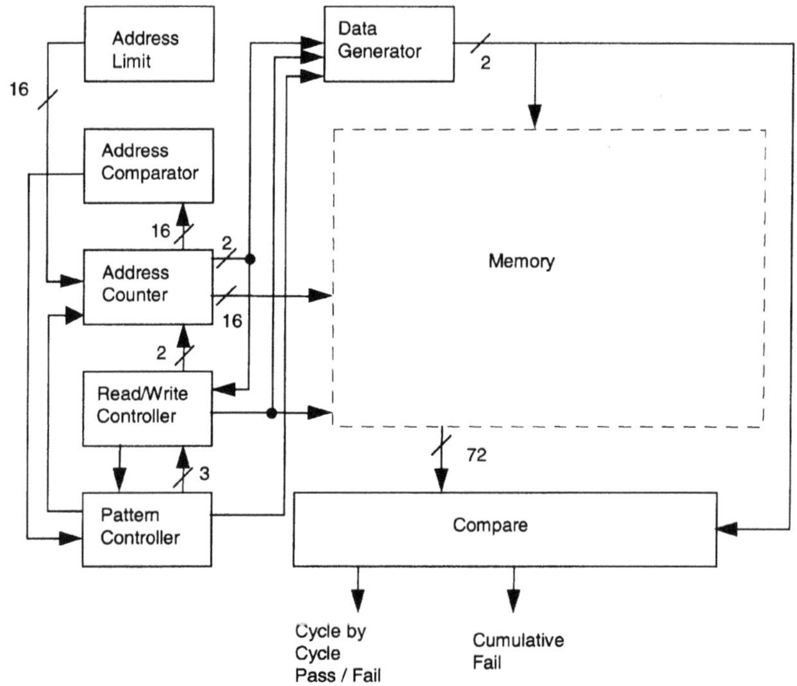

Figure 11-7. A BIST state machine block diagram.

The data generator interacts with the address counter, the read/write controller, and the pattern controller. These interactions allow the data generator to provide the memory with the correct data corresponding to the particular element of the particular test pattern.

All of these portions work together to provide stimuli to the memory under test. The memory then provides its outputs to the comparator or compressor at the memory output [169]. Neither a comparator nor a compressor is part of a state machine proper. The content of the comparator or compressor is dependent on the memory and whether or not it is defective. The comparator or compressor works with the remainder of the BIST sections to form a complete test solution.

Certain pattern elements have only a single operation per address, such as writing zeros at each address for the first element of the March C- pattern, as shown in Table 9-5. Other elements may have four operations per address, such as a R0, W1, R1, W1 sequence from the enhanced March C- pattern in Table 9-8 element two. Additionally, alternating ones and zeros may need to be generated, such as when a logical checkerboard is desired. Each such sequence can be handled by this type of state machine BIST.

5. PROGRAMMING AND STATE MACHINE BISTS

A BIST state machine follows a deterministic set of generated sequences but it is possible to program the BIST engine to execute modified patterns. A small number of latches can be included, which are initialized before the BIST is run and can enable rudimentary programming. This allows additional patterns to be defined which might include a group of just nine latches. The first latch could define whether to increment or decrement addresses for a memory sweep. There could be four bits defining a read or a write for four successive operations. Then there could be four bits defining the data to be read or written for those four operations. Thus a "R0, W1, R1, W1 ⇑ " could be defined with just nine programming latches in a BIST state machine, as shown in Figure 11-8.

Latch 1 Inc / Dec	Latch 2 R/W Cycle 1	Latch 3 R/W Cycle 2	Latch 4 R/W Cycle 3	Latch 5 R/W Cycle 4	Latch 6 Data Cycle 1	Latch 7 Data Cycle 2	Latch 8 Data Cycle 3	Latch 9 Data Cycle 4

Figure 11-8. Latches for simple programming in a state machine BIST.

Key point: A complex pattern suite is possible and most useful for memory test with a state machine BIST.

6. COMPLEX PATTERNS

A state machine BIST, due to its custom tailoring, can be utilized to generate very complex patterns. The sequence of loops and states required are generated in the BIST engine. One example of a complex pattern is shown in Table 11-5. This pattern is used in a memory with a very wide write port and a very narrow read port [170]. Unless a check is performed by writing a single "1" in the wide group of zeros, and then reading each of the locations, it is not possible to ensure that the decoder is defect free. Obviously, designing a state machine BIST engine to generate this type of pattern requires a highly skilled BIST designer, yet it is possible.

Table 11-5. Pseudo-code describing a complex state machine BIST pattern.

```
for i = 0..19 {
  write data_input(i) to a "1" with all other data_inputs at a "0" to row 0 ;
  for j = 0..19 {
    if (i==j){
      read "1" column i, read "1" column i ;
    } else {
      read "1" column i, read "0" column j ;
    }
  }
}
w0s to row 0
```

7. CONCLUSIONS

A state machine BIST generates a sequence of operations to provide the needed memory patterns. The state machine BIST can be simple and generate only a basic pattern or it can be complex, generating a suite of memory test patterns.

Still greater complexity is often needed, especially when considering multi-port memory [171,172] or CAM test patterns. For multi-port memories it should be recalled that multiple addresses need to be generated simultaneously [173]. These addresses may be at an additive value to the base address or one port's address may be incrementing while the other decrementing. For CAM, match line and other testing must be performed in addition to the base memory cell patterns. Thus, the BIST state machine capabilities must be vastly complex, when a thorough test pattern suite is utilized.

Chapter 12

Micro-Code BIST

Memory Self Test

"...if ye keep in memory what I preached unto you ..." I Corinthians 15:2

There are many advantages to a programmable memory BIST and a micro-coded BIST is the most programmable [174]. As a new technology is developed, new fault models can become evident during initial fabrication. The BIST test should be modified to include a pattern to find this new type of fault. A simple state machine BIST has only a predetermined pattern; modifying the pattern requires changing the BIST logic design. Even with a programmable state machine, such as that described at the end of chapter 11, the flexibility is quite limited.

Many memories need the flexibility of a micro-code BIST to ensure high test quality. A particularly challenging memory type, such as an embedded DRAM especially in a new process, can require many changes to a test pattern. Diagnostic debug can be aided by modifying a BIST pattern. Characterization of a new design can be facilitated by experimenting with various patterns. All of these tweaks, modifications, and even new patterns can be enabled through a micro-code BIST.

1. MICRO-CODE BIST STRUCTURE

A micro-code BIST contains many of the same features as a microprocessor, only on a smaller scale [175]. In some literature a micro-code BIST is referred to, instead, as a processor-based BIST. The components of the BIST processor are much more dedicated in purpose and therefore can be vastly simplified, as compared to a microprocessor. Even so, a micro-code BIST needs to have a BIST engine in addition to an instruction memory, where the definitions for the patterns are contained. Given the presence of instruction memory, a micro-code BIST is larger than a standard state machine BIST. Since a micro-code BIST is larger and since silicon area is critical, such a BIST should only be used when the extensive programmability is anticipated to provide considerable value. Figure 12-1 shows the primary components of a micro-code BIST.

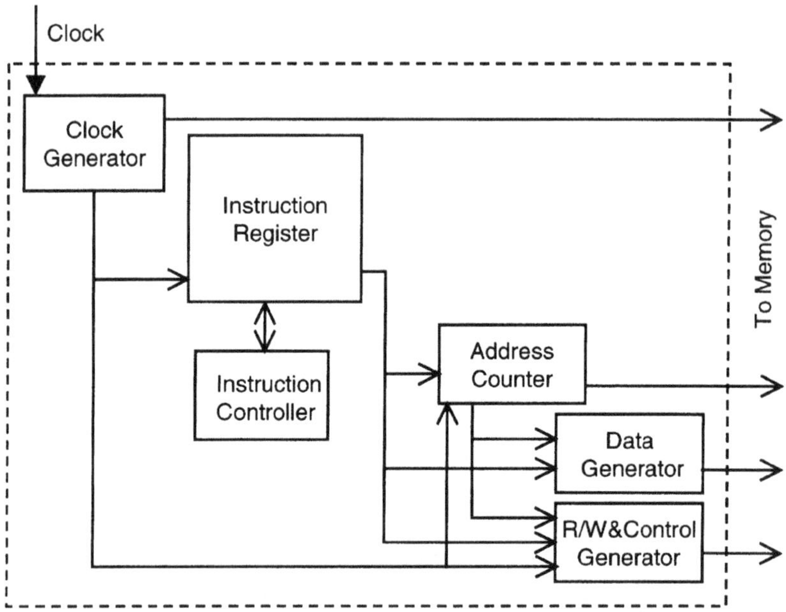

Figure 12-1. Micro-code BIST block diagram.

The clock generator supplies clocks to each portion of the BIST in addition, in this case, to supplying clocks to the memory during test. The instruction register supplies instructions to the address counter, the read/write and control generator, as well as the data generator. The instruction controller or sequencer detects when an address sweep has been completed and pulls a new instruction from the instruction register. The instruction does not define a complete pattern but rather defines the operations to be performed in a single address sweep.

Consider the example of "R0, W1, R1, W1 ⇓." The corresponding micro-code instruction tells the address counter to start at the maximum address. Further, it tells the address counter to decrement every fourth cycle. The data generator is told to generate a 0, 1, 1, 1 sequence to the memory data inputs and for the expect data to be compared at the output of the memory. The instruction also tells the read/write controller to generate a read, write, read, write sequence. If there had been other controls in the read/write control generator, such as bit-write controls or others, the instruction would have provided direction for these memory input portions as well.

2. MICRO-CODE INSTRUCTIONS

There are no standards to micro-code BIST instructions. A micro-code instruction word is highly honed to the particular memory structure under test. A multi-port memory BIST would be radically different from a BIST that handles several single port memories on a microprocessor chip. A BIST which fully tests a group of high performance multi-port memories would be different still.

An example micro-code BIST instruction word is given in Table 12-1. The first row indicates the operation while the second row indicates the bit position in the micro-code instruction word. The third row gives the number value assigned to each field in the instruction, which specifies the operation to be performed or the data value to be used. The first column gives the march element number. For example, bits 10 through 13 give data bits 0 through 3, indicating the read or write data values for the next four operations. The first bit in the word indicates that the instruction contained in this entry is valid. Normally test stops at the first invalid instruction that is reached [176]. The next two bits define the number of operations which are to be performed on a given address. The fourth bit defines whether the address counter is to increment or decrement through the address space. In this case a "0" indicates increment. The fifth bit tells the address counter whether the rows or columns are to be incremented most rapidly. If the bit is a "1", rows are rippled most often. The next four bits define the read/write operations for four cycles. Read, in this example, is a "0". For this micro-code BIST, four operations is the maximum number which can be performed on each address. If less than four operations per address are being specified, the remaining unused fields are “don't cares.” The next four bits define the data which is to be applied to the memory data inputs or expected at the memory data outputs, depending on whether a write or a read is performed. The last bit in this instruction defines if a checkerboard pattern is to be applied to the memory inputs. If the checkerboard bit is a "1" and the data bit is a "0" then a true checkerboard is to be applied. If the data bit is a "1" then an inverse checkerboard is to be applied. The polarity of any of these instruction bits can be inverted from a definition point of view. The convention just needs to be understood and agreed to by the BIST designer and the test engineer.

Table 12-1. Example micro-code BIST instruction word.

#	Valid	Operation Number		Inc/Dec	Row/Col	Read/Write				Data				Checkbd
	1	2	3	4	5	6	7	8	9	10	11	12	13	14
	1	0	1	1	1	0	1	2	3	0	1	2	3	1

Many more bits are possible in an instruction word. For a multi-port memory, the instruction needs to define which port or ports are being accessed. The information in the preceding example needs to be included for each port under test. Another bit can define whether these multi-port operations are to be performed serially or in parallel. If bit or byte-write controls exist, their input condition needs definition. The list goes on examining each of the memory's inputs to ensure that the test capability is complete.

The proper way to define a micro-code word is first to examine the memory design, determine the appropriate fault models, and determine the needed test patterns. Once the initial BIST patterns are defined and possible patterns anticipated, an optimal instruction word can be generated. The instruction word should include flexibility to implement patterns that might be needed based on new fault models from manufacturing defects.

Taking the example micro-code instruction word shown in Table 12-1, it would be helpful to illustrate what an instruction sequence might be for a March LR pattern, as shown in Table 9-9; the pattern is repeated here in Table 12-2 for clarity sake. The sequence of micro-code instructions to implement such a pattern is shown in Table 12-3. The first column indicates the march element number. The first word simply writes zeros throughout the entire array and therefore only one operation is defined per address. Since having zero operations per address is meaningless, a 00 entry in the second and third bits indicates one operation. Similarly a 11 would indicate four operations per address, as in march element 3. The X's indicate don't care positions that can be filled with either ones or zeros. The second word includes two operations, a read "0" and a write "1" in a decrementing order. Word number three corresponds to the walking part of the March LR pattern from element number three, a "R1, W0, R0, W1 ⇑" sequence. This pattern continues and concludes in element six with a read of ones incrementing through the memory array. The seventh word shows a zero in the valid column indicating the end of the test pattern. This type of a micro-code BIST can be used to program almost all of the patterns listed in chapter nine.

Table 12-2. Description of the March LR pattern.

1	W0 ⇕
2	R0, W1 ⇓
3	R1, W0, R0, W1 ⇑
4	R1, W0 ⇑
5	R0, W1, R1, W0 ⇑
6	R0 ⇑

Table 12-3. Micro-code instruction sequence for a March LR.

#	Valid	Operation Number		Inc/Dec	Row/Col	Read/Write				Data				Checkbd
	1	2	3	4	5	6	7	8	9	10	11	12	13	14
1	1	0	0	1	1	1	X	X	X	0	X	X	X	0
2	1	0	1	0	1	0	1	X	X	0	1	X	X	0
3	1	1	1	1	1	0	1	0	1	1	0	0	1	0
4	1	0	1	1	1	0	1	X	X	1	0	X	X	0
5	1	1	1	1	1	0	1	0	1	0	1	1	0	0
6	1	0	0	1	1	0	X	X	X	0	X	X	X	0
7	0	X	X	X	X	X	X	X	X	X	X	X	X	X

The micro-code instructions can be contained in a number of locations and in a number of types of memory. If all of the BIST test patterns are planned before the design is complete and if no changes in test patterns are ever desired then a simple ROM can contain all of the instructions. If, on the other hand, total flexibility on all of the patterns is desired at all points then a set of registers can be programmed for each execution of the BIST. These registers can be in the form of scannable latches, a register array, or even an SRAM. Generally, it is not desired to use an SRAM to contain instructions for a memory BIST since the SRAM must, in turn, be thoroughly tested before running the BIST. A good programmable alternative is to include a base set of patterns in the ROM and be able to augment those with instructions that can be loaded when the BIST is run [177]. These instructions can be loaded by scanning, through a test access port, or by some microprocessor bus on chip. If a ROM is utilized, the ROM can even reside off-chip [178], although some of the benefits of having a BIST on chip are lost.

3. LOOPING AND BRANCHING

The micro-code BIST example shown in Table 12-3 illustrates a simple BIST structure which executes instruction after instruction until the test is complete. A more complex BIST structure allows branching and looping. The preceding example does show some primitive branching and looping. The looping included in this previous BIST allowed a march element to be repeated on each address. It also allowed between one and four instructions to be executed per address. These capabilities are much expanded in typical micro-code BISTs.

A micro-code instruction word can be broken up into a pre-instruction section and a series of sub-instructions. In the previous example the pre-instruction portion would be the first three bits. These bits detailed that the instruction was valid, i.e. that the BIST test was not complete, and the number of operations to be performed on each address. Since up to four operations could be described, the remainder of the micro-code instruction word could have been broken up into four sub-instructions.

A more complex pre-instruction can contain the information already described plus it can also include information on branching and looping. In the case of the micro-code being stored partially in ROM and partially in programmable registers, branching can be quite useful [179]. It is possible to store a base set of patterns in the ROM with a branch out of the ROM after each individual pattern. In this manner, the programmable register can contain a series of branches. These branches tell which pattern in ROM to execute and in which order. If a pattern already stored in ROM is to be added, one more branch instruction is included in the programmable register. Other branching can be included as well, similar to the types of branching utilized in any software program.

Looping can provide a powerful capability in BIST test patterns. The simplest loop would be executing the same instruction over and over again. It can be useful in a characterization mode to repeat a test infinitely, allowing an e-beam prober to collect an electrical waveform over vast numbers of cycles. The instruction can be repeated for the whole address space or, more likely, on a limited address space of one or two memory locations. Looping can also be used to execute a sequence of operations to facilitate a complex memory test pattern. Some BISTs have been designed to allow nested loops, with the nesting being up to a depth of seven.

Figure 12-2 illustrates some of the interactions inside a micro-code BIST controller which facilitates branching and looping. The sub-instructions are seen going to the address, data, and read/write control generators. The pre-instruction goes to the instruction dispatch block which provides looping and branch control to the instruction counter and pointer block. This block, in turn determines, for the micro-code instruction storage block, which instruction word to access next. The dotted line indicates that the instruction storage can be partly ROM and partly programmable register file.

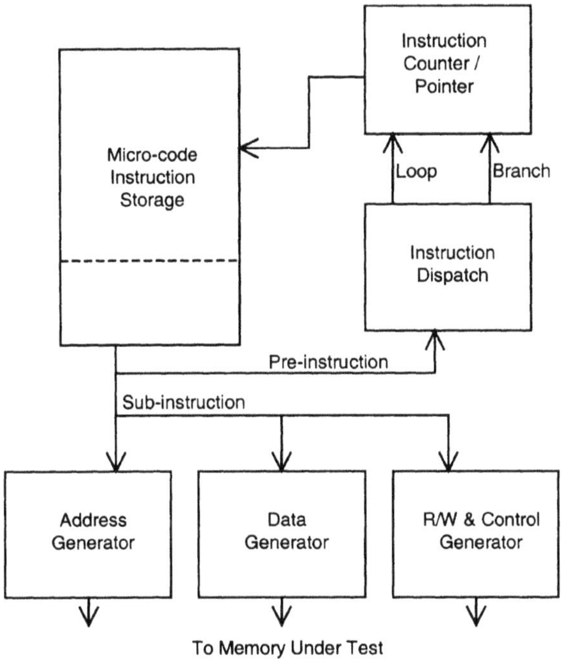

Figure 12-2. Micro-code instruction control.

4. USING A MICRO-CODED MEMORY BIST

A micro-code BIST can be very helpful when generating patterns for an embedded DRAM (eDRAM), especially in a new technology [180]. Since there are more analog effects in a DRAM there is more need to modify the pattern set during initial characterization and test development in order to detect subtle defects. Another cause for significant pattern modification arises since DRAM testing requires verifying the specification, i.e. checking the cell retention time, rather than just looking for manufacturing defects. Signal margin must also be tested to ensure high test coverage in DRAM cell arrays. It is the role of the characterization and test development team, during initial hardware verification, to thoroughly determine the correct set of patterns for this type of memory. It is nearly impossible to predict all of these patterns at the time of initial design submission.

In one example, a 34-bit instruction was utilized for advanced eDRAM BIST testing. An average of eight of these instructions were fetched to compose a complete test pattern. A total of 256 instructions were implemented without reloading the BIST. Reloading would enable even more patterns to be executed. Given the complexity of eDRAMs, 14 patterns were implemented to define a complete test. Sixteen different types of conditional jumps were included in the design to enable a variety of patterns.

Key point: A micro-code BIST is the most flexible memory self-test method.

With all of the possible permutations to generate various patterns, looking at the sequence of ones and zeros can become very tedious and error prone [181]. Therefore a human-readable set of codes was generated for a number of memory BISTs. These codes are compiled on an ATE or other system, automatically verified for accuracy, and provided to the embedded BIST. In this manner, the correct complex patterns are easily generated and applied by the BIST.

It should be noted that a programmable pattern can be generated which will always fail or worse, will always pass. For example, it is possible for one instruction to write zeros into a memory. On the next instruction, it is possible to tell the BIST to read ones. This pattern should fail on every address. It is the test engineer's responsibility to ensure that the pattern is indeed valid for a defect free memory and detects any defective memories.

Some patterns are not conducive to be implemented by a micro-code BIST engine. Very convoluted patterns, such as address decoder patterns, require sufficient address jumps that implementing these with a straight micro-code BIST would be cumbersome. In this case a marriage of a state machine BIST and a micro-code BIST is optimal [182]. The micro-code BIST is the primary controller but, with a specified micro-code instruction, a secondary state machine can temporarily take over to generate the needed set of patterns. For fixed highly customized patterns, a state machine BIST would be most optimal. By using a small dedicated state machine to generate the unique patterns, with the state machine being subservient to the micro-code BIST, a highly programmable yet very tailored combination of patterns can be generated.

Using a micro-code/state machine BIST combination requires a special micro-code instruction to enable the state machine operation. Other special purpose micro-code instructions can be included to enable different test types. One example is an address specific micro-code test. In this case a micro-code instruction contains an address that is loaded into the address generator. A sub-instruction can then operate on just that address. Other types of BIST micro-code instructions can be developed as needed for specific memory topologies.

Other micro-code instruction bits can enable added design-for-test (DFT) features. One example is a shadow-write capability that aids in detection of bit-line shorts. A micro-code bit can be set for a given instruction that turns on the shadow-write feature for several march elements. Other DFT features, as will be described in chapter 14 can similarly be enabled for specific patterns with dedicated micro-code program bits.

A micro-code BIST can be used in both a manufacturing and in a system environment. Since the micro-code BIST is programmable, different patterns can be utilized in each. In system applications, it is not desirable to perform a complex initialization of a BIST. Instead, a base set of patterns should be programmed into the micro-code BIST so that on power up a simple reset can enable the BIST to execute these patterns. If scanned registers are utilized for programming the micro-code BIST, a flush of zeros can initialize the BIST to the proper setting for the base patterns. Inverters need to be placed at the appropriate locations in the scan chain to facilitate the proper initialization but the BIST designer can include these. If another type of micro-code storage memory is being utilized it can be reset in some appropriate fashion to generate the base pattern set. During manufacturing test, the micro-code BIST can be programmed to do a more involved or special test. Similarly, during initial design characterization, the micro-code BIST can be programmed to do any number of tests to aid the characterization engineer.

5. CONCLUSIONS

A micro-code programmable BIST is the most flexible of self-test structures. The memory test patterns can easily be modified based on new fault modeling or to assist in the characterization of a new memory design. A micro-code BIST needs to be tailored to the memory design being tested and therefore a wide variety of micro-code instruction word styles exist. Many unique patterns and DFT features can be programmed with a micro-coded memory BIST; some of those items include pause delay duration for retention testing or the number of times a given loop is executed. Each of these features enables a micro-code BIST to be highly flexible to support very challenging memory testing.

Chapter 13

BIST and Redundancy

Memory Self Test

"*... beside them all orderly in three rows ...*" – Homer's *The Iliad*

Redundancy is very critical to generating yield. Memories are very dense structures which allow ample opportunity for defects to impact their operation. Memories are much denser than any logic, with long runs of metallization with minimum distances separating one wire from the next. The width of these wires is again minimum for a given technology. The diffusions are packed in as tightly as possible with polysilicon, well shapes, and diffusions all at minimum distances. Ground rules are defined to ensure good manufacturability and printability of structures. Since many bits need to be packed as tightly as possible, ground rules are waived in the memory array region. The fabricator can produce these new dimensions which match the ground rule waivers but smaller defects, which are more plentiful than larger ones, can now cause a fail.

Defects are a part of any fabrication process and even though minute, these defects can cause havoc in a memory circuit. A defect can easily open a wire or short two diffusions, polysilicon shapes, or metal wires together. Just as life is not perfect, memories must be able to exist in an imperfect world, i.e. in the presence of defects. Redundancy enables this to happen.

Most memories of any significant size have redundancy and even smaller memories may require it. Even if the individual memories are small, with sufficient numbers of memories the total number of bits becomes large. The requirement for redundancy is driven by the total number of bits on a chip, not by the size of each individual memory.

For larger memories it is easy to have many millions of bits, just in redundancy. These extra bits are in the form of redundant rows, redundant I/O, redundant columns, and redundant blocks. Whereas smaller memories need only a single type of redundancy, larger memories need multiple redundancy dimensions.

1. REPLACE, NOT REPAIR

When a memory fails, it usually has a bad single cell, bad row, bad column, or bad cluster. The dominant fail mode is single cell fails, followed by paired cell failures, and then by row or column failures [183]. Memories, after test, can be good, bad, or fixable. Sometimes memories, instead of being called fixable are said to be repairable. Being repairable or fixable carries a connotation that the bad item is going to be repaired and made good again. Memories don't work like that.

If a memory element is bad it is not repaired but is instead replaced. There are extra memory elements, such as rows, columns, or blocks, on a memory. When an defective element is detected, that element is ignored and one of the spare elements replaces it. When someone says the word "repair" it sounds as if the memory is being fixed with a microscopic soldering iron. This is not the case and the bad piece of the memory is steered around and a substitute piece is used in its place. Although the terms repair and replace are used interchangeably, the correct meaning should be understood by the designers and test engineers.

2. REDUNDANCY TYPES

Since most fails on memories are single cell fails, it would be fine to be able to repair a single cell at a time. The overhead to repair a single cell at a time is too large to be practical. Identifying a single cell to be replaced means storing the failing row address, column address, and I/O bit. On each read a compare would have to be performed on all three portions of this address against the stored failing location. If the failing location is accessed, the redundant single cell would then be supplied in place of its failing counterpart. The time to do this substitution, not to mention the amount of redundant address storage, would be very prohibitive. Instead of individual cells being replaced, rows, columns or even whole blocks or sub-arrays are replaced. Figure 13-1 shows a memory with redundant rows and I/O replacing failing locations. The redundant rows and I/O are allocated on a quadrant basis.

When a row is replaced, that row spans all columns and sub-arrays in a quadrant. When an I/O is replaced, that I/O spans all rows within a sub-array. Multiple single cell fails in a dimension can thereby be replaced at once. In some cases row pair group is replaced. A defect that lands in a Vdd contact or in a bit-line contact causes a vertical pair of cells to fail. By replacing a row pair, these type fails are covered without having to use two independent redundant rows to replace adjacent failing bits.

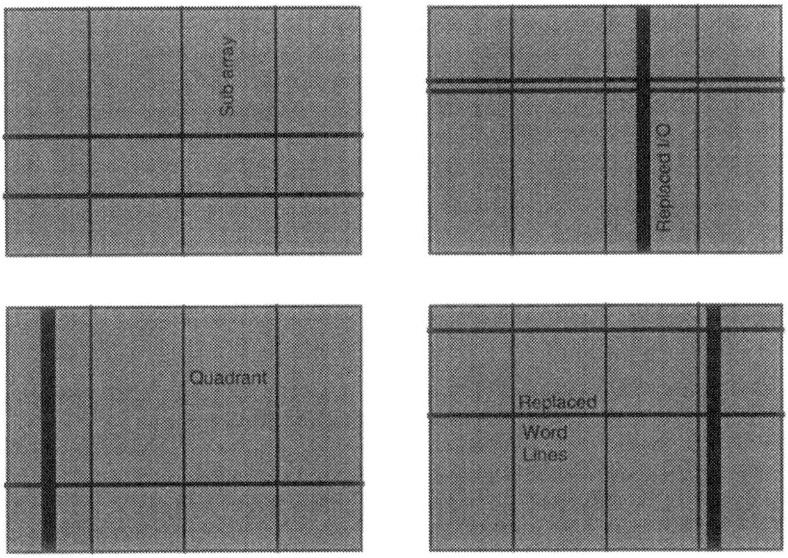

Figure 13-1. Memory utilizing redundant rows and I/O.

Column redundancy involves replacing a column throughout a sub-array. This column requires at least a single bit line and most often a bit-line pair to be replaced. Determining which column to replace means determining the I/O that failed along with the specific column address. Since the overhead on replacing columns is large, often an entire I/O is replaced [184]. Replacing an I/O covers a group of eight, 16, or more columns, depending on the specific address decode factor. By replacing an I/O, the pre-charge, write driver, sense amplifier, and column multiplexer circuits are all replaced [185]. Therefore many non-single cell type defects can be replaced including problems with the listed circuitry as well as bit line shorts. Even a cluster of fails can be neatly maneuvered around. Figure 13-2 shows a group of single cell fails that are on different word lines and different column addresses. With redundant I/O replacement, this type of fail can be easily replaced, given an adequate redundancy calculation.

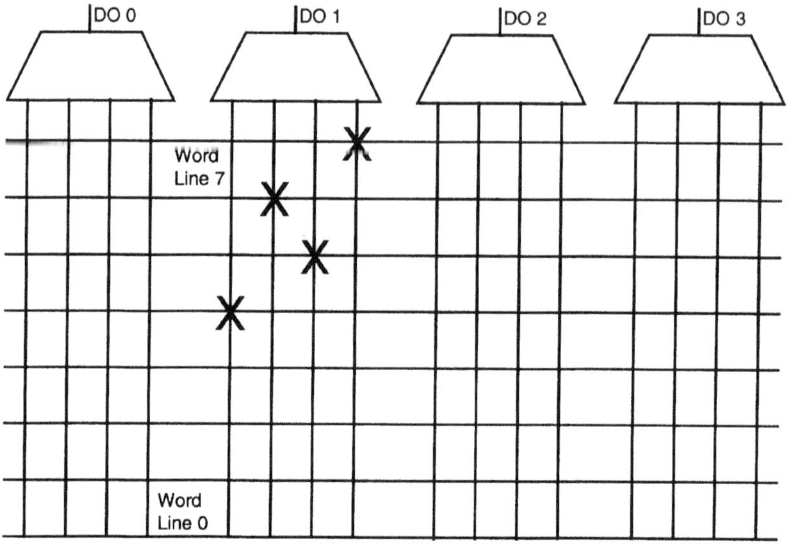

Figure 13-2. Failing I/O which should be replaced.

If a memory has a 16 to 1 column decode then 16-bit line pairs are all replaced simultaneously allowing for repair of large defects or numerous small ones. Figure 13-3 shows a redundant I/O replacement scheme with a four-to-one column address decode. Since a defect on the third bit line of data number one (D1) exists, the entire D1 I/O is bypassed. This method of steering allows any I/O to be replaced along with the columns, sense amplifiers, and associated circuitry [186,187].

Certain fails which should not be repaired, including defects which cannot be worked around or which may propagate over time. Most decoder failures should not be fixed as is the case with high current fails. Even though a failing location is replaced, it can still draw current. When developing a BIST and using it to perform a redundancy calculation, careful consideration should be taken to identify failures which should not be replaced. A fail setting should indicate that no repair is possible rather than storing the redundancy fix information in BIST.

The amount of redundancy on a memory is dependent on the type and size of the memory. Redundancy for certain types of memory can be more challenging than others. An example is a content addressable memory which needs not only to store information but perform match compares and mask functions as well. A ternary CAM has even more challenges in that it must store three states and be able to mask on a per bit basis for each memory entry. The compare circuitry, as well as the memory cells, needs to be replaced for this type of memory.

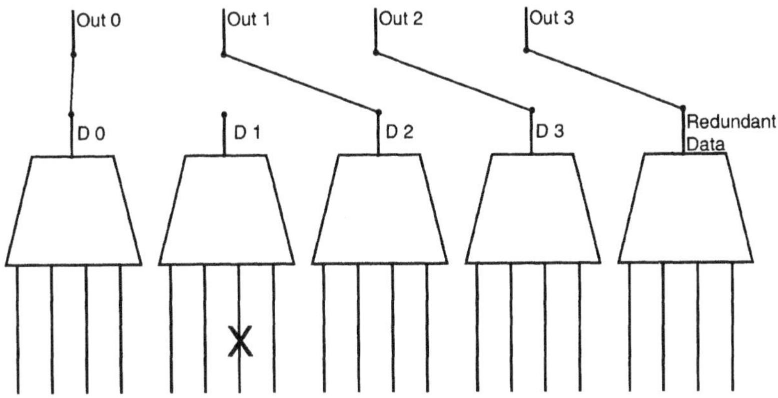

Figure 13-3. Redundant I/O steering.

For embedded memories, the first type of redundancy implemented is typically row. When more redundancy is needed, an extra I/O is provided. When even larger memories are to be implemented, a redundancy on column basis can be added. It is not just the amount of redundancy, in terms of bits that is key, it is the form factor and the granularity. For instance, as already stated a redundant row pair can fix a vertical cell pair but it cannot fix two vertically separated single cell fails. Thus redundancy is key and the type of redundancy is crucial as well. The right amount and type needs to be implemented in order to get sufficient chip yield.

3. HARD AND SOFT REDUNDANCY

Once the type of redundant element has been defined for a memory the next step is to determine how redundancy is invoked. It can be either of the soft redundancy or the hard redundancy variety.

A hard redundancy implementation utilizes some form of fuses to store the information for memory element replacement. The fuses can be laser fuses, electrical fuses, or even EPROM memory [188]. If a laser fuse is utilized, the correct redundancy calculation is performed and the information is uploaded off chip. This information is communicated to a laser fuser which then opens certain on-chip fuses with a laser cut. A laser fuse needs to have an opening exposed to the top passivation of the chip so that the laser light can reach and cut the on-chip fuse level. If electrical fuses are utilized then the appropriate fusing information needs to be fed across the chip to open the appropriate fuses for the correct redundancy implementation [189]. An anti-fuse can be used in some technology where applying stimulus actually shorts a path rather than the typical opening of a path with normal fuse structures [190]. EPROM type memory can store the appropriate redundancy implementation and can be modified if an EEPROM or some other modifiable non-volatile memory is utilized [191]. By storing redundancy information in this type of structure it can be updated at various levels of testing.

Soft redundancy is calculated at each chip power on. Through a power-on-reset of similar invocation, a BIST test is performed and redundancy is calculated. This information is held in latches which directly tells the memory which elements to replace. The soft redundancy calculation can be enhanced through occasional transparent BIST exercises that update the stored redundancy calculation. One of the disadvantages of soft redundancy is the existence of marginal fails. A subtle memory defect can cause failing operation at one temperature and not at another. The concern is that a subtle defect will pass test at room temperature and then as the use condition temperature is reached, the memory will start to fail. There are not many of these type defects but initial manufacturing test must ensure that marginal defects are not shipped, even if the redundancy can work around them.

A combination of hard and soft redundancy can be utilized to enhance a memory's fault tolerance. With this method, a certain amount of redundancy is implemented by fusing means and the remainder can be implemented with soft redundancy calculations while the system is in the field. Using BIST to do the redundancy calculation in the system is often referred to as built-in self-repair or BISR [192].

4. CHALLENGES IN BIST AND REDUNDANCY

There are many challenges presented to a BIST in performing the proper redundancy calculation. The perfect redundancy calculation is difficult in that it is an NP complete problem, when multiple dimensions of redundancy are possible. Since a single cell fail can be replaced by a redundant row, column, I/O, or block, it shouldn't matter which type of redundancy is utilized, if that is the only fail on the memory. The reason that multiple dimensions of redundancy are designed into the memory and the reason that many redundant elements are available is that many defects can impact a single memory. Determining which redundancy dimension to invoke to repair which fail is crucial since the wrong decision yields an unfixable chip.

ATE testers store all of the information on the failing bits for a memory before deciding which redundant elements to utilize where. This amount of information cannot be stored in a BIST environment since the same amount of storage is required as the memory under test. BIST must perform on-the-fly redundancy calculations before all of the information is found on the fails for a given memory. That means that a very intelligent BIST with good redundancy allocation logic must be used to do the appropriate calculations. Some ATE manufacturers have argued that BIST should not perform the redundancy calculation and that the needed information should be sent off chip for the needed calculation [193]. While this approach sounds acceptable at first glance, the possible number of memories on a chip quickly makes this approach unpalatable.

The number of memories that need repair in a BIST environment is the next challenge that needs discussion. In light of hundreds, and shortly thousands, of individual memories on a chip, getting failing information off chip would create an unbearable test time burden. Instead the BIST needs to handle the memory testing and the redundancy calculation for all of the memories with redundancy. A single BIST can easily test multiple memories, however, for each memory with redundancy separate redundancy-allocation logic needs to exist. Since the memories are tested in parallel and fails can happen on each of the memories under test, the appropriate redundancy implementation needs to be calculated for each memory, all at the same time. An alternative is to test each of the memories serially and use a single redundancy allocation logic unit to perform each of the calculations one memory at a time. The decision is one of prioritizing silicon area versus test time. Multiple redundancy-allocation logic units require more area on the chip whereas performing each of the memory tests serially takes more time. Different applications have different priorities. For a high volume part, test time would be paramount. For a unique chip with low volumes, silicon area might be paramount. These factors drive decisions on the redundancy implementation and calculation.

Other challenges in redundancy calculations for BIST involve the complexity of the memory under test. For a multi-port memory, with multiple operations going on simultaneously, it may be difficult to determine which port and which address are actually defective [194]. Careful BIST pattern selection is needed to ensure that the failing location is flagged for replacement. Similarly, CAMs and other complex memories need to have their cells and the associated logic all replaced with the appropriate redundancy.

A last challenge in redundancy and BIST is dealing with defective redundant memory elements. It is easy to see that with a large amount of redundancy it is possible to have defects land in these areas. After all of the effort to perform a BIST test and a redundancy calculation, it is possible to have a failing memory element replaced with another failing memory element. This result certainly doesn't enhance yield. To work around this type of problem it is helpful, although not trivial, to test the redundant elements prior to selecting the specific redundancy invocation.

5. THE REDUNDANCY CALCULATION

If a single dimension of redundancy exists and a fail is encountered, the redundancy calculation is trivial. For instance, when a fail is encountered a pass/fail flag is sent to the redundancy calculation logic by the comparator at the output of the memory under test. The failing row is stored in a failed address register, as shown in Figure 13-4. The row address is stored and a valid bit is set in the first entry. When another fail is encountered, a comparison occurs looking for a match with the row address previously stored. If there is a match the new fail is ignored. If no match exists, the new row address is stored in the next entry and its valid bit is set. This continues until test is complete or until all of the failed address entries are full and a new fail is detected which hasn't been stored. In this case the memory is not fixable. Otherwise the failing locations can be replaced by redundant elements as defined by the addresses stored in the failed address register.

When multiple dimensions exist, the calculation becomes much more interesting [195]. Figure 13-5 shows an example of five failing bits in an eight by eight bit memory, where the failing bits are denoted by a cell with an "F" in it. Randomly invoking redundant elements as each fail is detected, clearly will not suffice. If two redundant columns and a redundant row are available to repair the memory, it would be easy to allocate a redundant row to the pair of fails in row three, as shown in Figure 13-6. In this the other failing cells could not be repaired. If instead the fails in columns three and six are replaced with redundant columns, the fail in row eight can be replaced by the redundant row, allowing full repair. This repair invocation is shown in Figure 13-7.

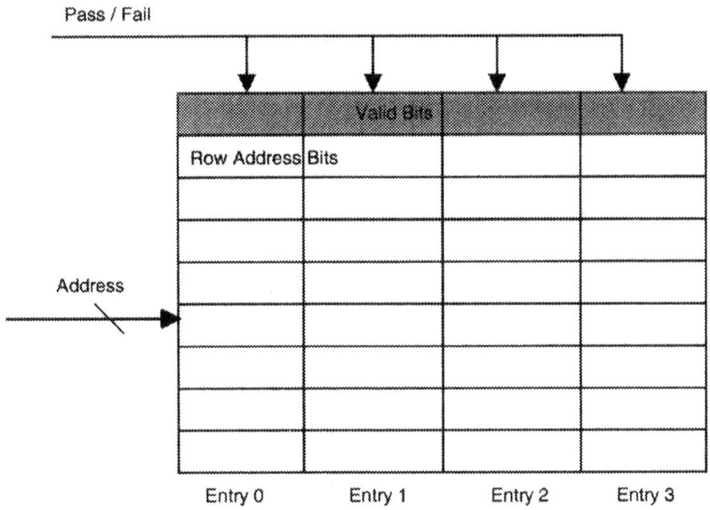

Figure 13-4. Redundant row failed address register.

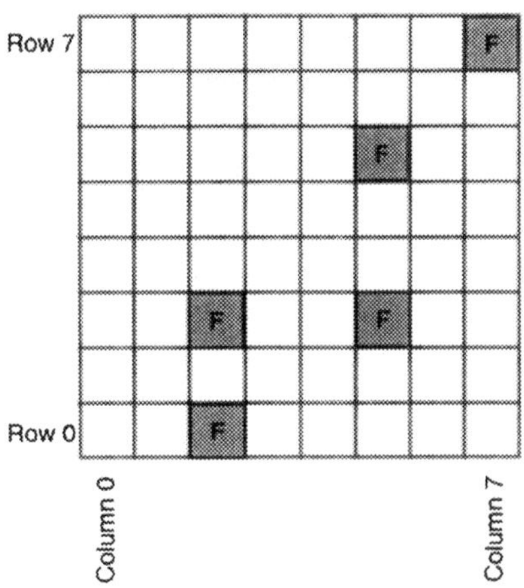

Figure 13-5. Memory with five failing bits.

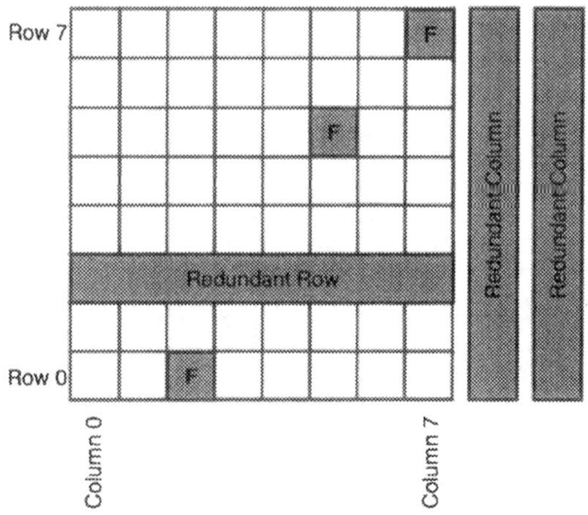

Figure 13-6. Row three is allocated a redundant row. Result is in an unfixable memory.

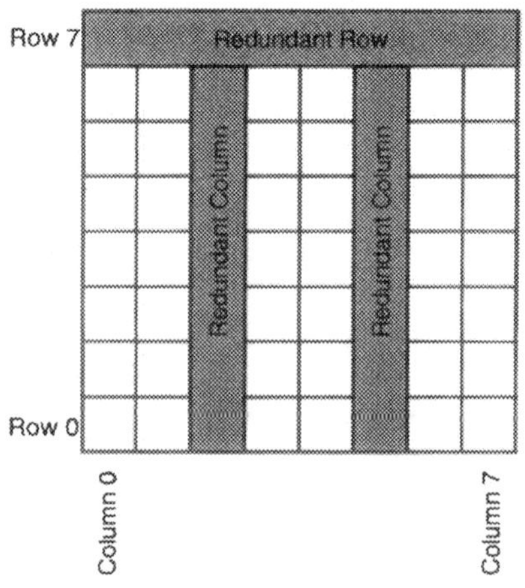

Figure 13-7. Memory fixed with correct redundancy allocation.

Similar challenges can be found with various other multiple dimension redundancy schemes. It is important that the BIST correctly implement redundancy to avoid unnecessarily throwing away repairable chips. One method for calculating when to replace a column and when to replace a row employs two pass testing of the memory with BIST. On the first pass through, the number of fails along rows and along columns is counted. When a specified number is exceeded in either dimension a "must-fix" row or column is defined. Certainly, if there are four redundant rows and five fails are detected along a column, a must-fix column selection can be made. Other values can be chosen to determine when the must-fix selection should be made, based on the specific memory topology and the fabrication technology being utilized.

Key point: Correct redundancy calculation by BIST enables high yield.

Once the must-fix determinations are made and all of the larger defects are marked for repair, a second pass test is performed through the memory to allocate the remaining redundancy to the sparse failures [196]. Some pathological fail cases can occur when a memory is fixable but the wrong implementation is chosen. As a result the memory is deemed unfixable. Since these cases are rare, the memory yield is significantly enhanced, although not perfectly.

Even with a good redundancy calculation method, the best BIST implementation also cannot assure a repair of 100% of the fixable memories. Proper BIST implementation of a redundancy calculation, though, can provide significant yield enhancement and can also assure that no failing memories are shipped to the field.

6. CONCLUSIONS

Redundancy is a key enabler to successful chip yield. It allows imperfect memories to be repaired for full customer functionality. A BIST needs to identify failures and determine the optimal redundancy repair scheme. With multiple dimensions of redundancy, this calculation is non-trivial and requires careful consideration of the specific memory topology under test.

Chapter 14

Design For Test and BIST

Memory Self Test

"I have these memories ..." – from the movie *The Matrix*

Design-for-test techniques are aids to enable detection of defects. A design-for-test technique involves modifying a memory design to make it testable. In the world of design for test, the emphasis is not on test but rather on design. How can circuits, and more particularly memories, be designed so that defects can be detected? This chapter details some design-for-test (DFT) techniques that are very helpful for detecting subtle defects.

Many defects are subtle and are difficult to detect in a normal environment let alone a test environment. These defects could easily be missed and a faulty memory allowed to pass test and be shipped to a customer, where it could fail intermittently. It is very beneficial to test a little more strenuously than a customer's requirement to ensure that there won't be problems in the field. For these reasons DFT techniques are included to test circuits beyond their normal limits to identify defects that could fail in the field. Several key DFT techniques for memory are detailed in this chapter but many others exist.

Memories are dominated by the analog measurement of a "1" or a "0" state in the cells, involving a slight shift in the bit line potential. The slight shift in the bit line potential that distinguishes between a "1" and a "0" means that subtle defects can impact this operation. Similarly, test techniques can be developed which utilize the dependency on small changes in bit line potentials to enhance defect detection. Throughout this chapter techniques are describes which push voltage potentials in the opposite directions from the norm. This exacerbates defective weaknesses to the point of detectability.

1. WEAK WRITE TEST MODE

Normally an SRAM bit-line pair has both lines pre-charged to a Vdd condition. During a read, one bit line discharges a small amount. On a write operation, one bit line is driven to ground while the other is maintained at Vdd. Midlevel potentials are never maintained on a bit line during normal operation. In the weak write test mode (WWTM) [197], bit lines are held at abnormal values while the word line for each address is driven high, enabling a weak write of each cell to the opposite value from that stored in the cell. A defect-free cell maintains its value while a defective cell is overwritten.

Example circuitry for performing the weak write test mode is shown in Figure 14-1. During a weak write of a "0", the WR0 line is driven high whereas during a weak write of a "1", the WR1 line is driven high. The weak write time is longer than a normal read or write and can be on the order of 50 ns. The sizing of the weak write FETs shown in Figure 14-1 must be carefully chosen so that a good cell is not overwritten while a defective cell is detected as failing. All process, temperature, and voltage corners must be considered in selecting the device sizes.

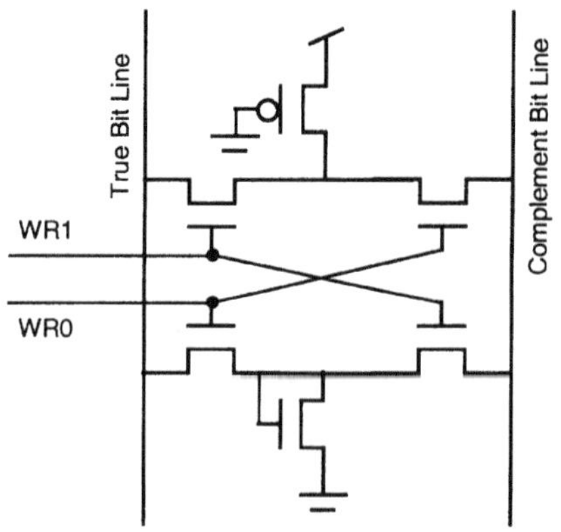

Figure 14-1. Schematic of weak write test circuit.

Weak write testing facilitates finding defects in the pull-up path of an SRAM cell. These defects can be on either the source or the drain side of the pull-up PFETs. If a defect is on the drain side, the defect affects operation asymmetrically. If the defect is on the source side and both cell PFETs share the same Vdd contact, operation can be impacted symmetrically. Both asymmetric and symmetric defects can be detected using the weak write test mode.

Defects have been known to exist in the pull-up path for some time but have been hard to detect since they affect retention and cell stability issues. Defects in the pull-down path were easy to detect since the bit lines were pre-charged to Vdd and a double read test normally detects these. Previous test techniques to detect defects in the pull-up path included a pause test or a bump test. Both of these are considered passive tests whereas the weak write test mode is considered an active test since it actively goes after detecting defects in the pull-up path. Additionally, the weak write test mode requires less test time than a typical pause type of retention test making it more attractive from a test time cost as well. The only penalty is the addition of six FETs on each column and the associated control signals.

2. BIT LINE CONTACT RESISTANCE

While the weak write test mode facilitates test of defects in the cell pull-up path, there are other sites where defects can impact operation. One such common location is in the bit line contact, between a cell and a bit line [198]. If a bit line contact is defectively resistive, it is hard to detect during normal testing but can impact operation in the field. A photograph of a resistive bit line contact is shown in Figure 14-2. The contact on the left is defective while the contact on the right is defect free. A highly resistive contact degrades read performance but not to the point of failure. Since signal margin is decreased and robustness is lost but functional operation can be maintained during manufacturing test.

An increase in the amount of bit line contact resistance actually impacts the write operation in a greater way than is seen during a read operation. Since the bit line needs to overcome the cell pull-up PFET, a higher than normal bit line contact resistance diminishes the drive capability to write the cell to the opposite state. Transfer devices, due to their small size, are already highly resistive in the on state thereby making an elevated bit line contact resistance more difficult to detect. Given these challenges, the addition of a DFT circuit can help identify defective bit line contacts.

The design modification of an added cell along a bit line can facilitate test. The added cell is designed to be weaker than a normal one and such that a normal cell could be used to overwrite this new DFT cell. To accomplish bit line contact resistance testing, a normal cell and the DFT cell are written to opposite states, through normal writing techniques. Then both the word line for the normal cell and the word line for the DFT cell are activated. In a defect-free condition, the DFT cell changes state to match the state of the normal cell. If the normal cell has a highly resistive bit line contact then each cell will remain in their preceding state. Since the DFT cell did not change state, a defective bit line contact resistance is detected on the normal cell.

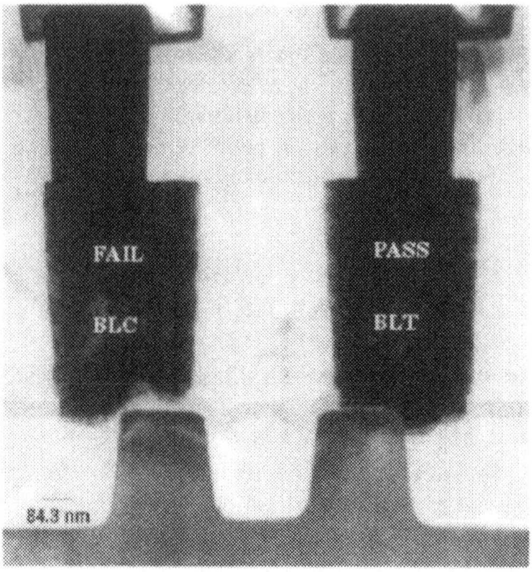

Figure 14-2. Photograph of resistive bit line contact.

For area optimization, most memory designs have a vertical pair of cells sharing a single bit line contact. When two cells share a bit line contact, the best bit line contact resistance test is accomplished by activating both word lines for the vertical pair of cells. Since both transfer devices are on as well as both pull-down devices for the vertical cell pair, any elevated bit line contact resistance is more easily detected. The DFT cell needs to be sized to be slightly smaller than twice the size of a normal cell, since it is fighting against a vertical cell pair.

The design modifications to enable this bit line contact resistance testing include the addition of the DFT cell as well as changes to the word line decoder drivers to allow multiple word lines to be activated simultaneously during test mode. Interestingly, a defective bit line contact resistance cannot be detected by a weak write test mode since the higher contact resistance actually makes the cell appear more stable. An alternative to the bit line contact resistance test includes shortening the word line up time during a write; the sensitivity to defective contact resistance is far less, though.

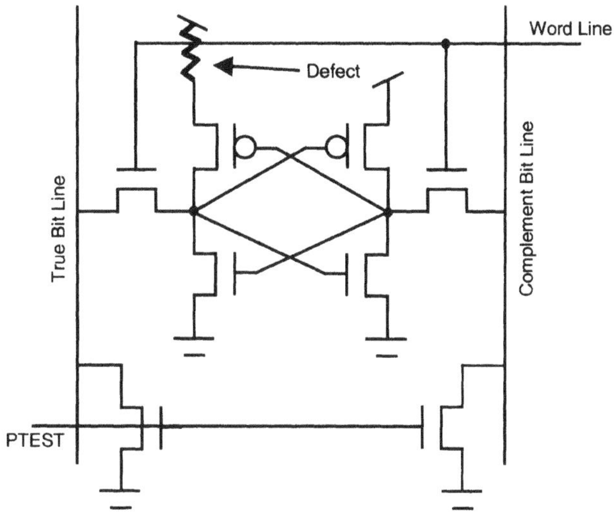

Figure 14-3. Defective pull-up path detected by PTEST.

3. PFET TEST

The pull-up path of a memory cell is difficult to test for resistive defects, as already stated. While one alternative for identifying resistive defects on the source or drain of a pull-up PFET utilizes the weak write test mode, another alternative exists. A PFET-test mode or PTEST utilizes a pair of NFETs for each bit-line pair in an SRAM [199]. The two NFETs are sized so that they can only weakly pull down the bit lines and both NFETs are activated at the same time. Each data type needs to be written into the memory array. Then the pre-charge is removed from the bit lines and the PTEST NFETs are activated, weakly discharging the bit lines. Each address is then accessed by bringing the word line high for its normal read or write duration. If a sufficiently resistive defect exists in the pull-up path of any cell, that cell will flip and be overwritten by the opposite data type. A subsequent read operation is performed to detect any cells which are in an erroneous state. Figure 14-3 shows an SRAM cell with a defect in the true pull-up path along with a pair of PTEST NFETs on the bit lines. When a "1" is stored in the cell and the PTEST is implemented, the weak NFET causes the cell to change states. One positive aspect of this DFT circuitry is that only two devices are required per column.

4. SHADOW WRITE AND SHADOW READ

Multi-port memories can have certain defective interactions which are hard to detect. These defects may be rather gross in nature and yet be hard to detect, due to the masking effects of multi-port operation. To enable better testing in multi-port memories, shadow-read and shadow-write operations were developed [200,201]. These are modifications of the normal read and write operations to facilitate test.

One type of shadow write is helpful for finding shorts between adjacent bit lines. In single-port memories, a short between adjacent bit lines is rather trivial to detect. A full bit-line potential swing, which is accomplished on a write, is severely impacted if the bit line is shorted to its neighbor. In a multi-port memory, one or more of the ports may have read-only utility. A read is accomplished by only having a small swing in bit-line potential. If a short exists to a neighboring bit line, which is also pre-charged high, it could easily go unnoticed since the potentials of the neighboring shorted lines are so similar. To detect shorts between adjacent lines they need to be at opposite potentials. A shadow write enables adjacent bit-line pairs to be at opposite states, even if each pair of bit lines is for read-only ports. A DFT feature is included to ground alternating bit-line pairs. If an A and a B read port exist then the A bit-line pair is grounded while the B port is read for each data state. Example DFT circuitry is shown in Figure 14-4. Any short between A and B bit lines would be detected. Next the B bit-line pair is grounded while each data type is read from port A. The shadow write DFT feature forces neighboring bit lines into states similar to those seen with a full writes, thereby detecting defects which would otherwise be missed.

A shadow read is a modification of a normal read operation to facilitate testing for other possible multi-port memory defects. On one port, a normal march-type pattern is exercised while on all other memory ports a read is performed. The result of the read on the other ports is ignored. By performing reads without looking at the result, certain inter-port faults are activated. These activated faults are detected by subsequent march operations. This is a DFT feature that is at the boundary of the memory and BIST whereas the DFT features already discussed are modifications in the memory design itself.

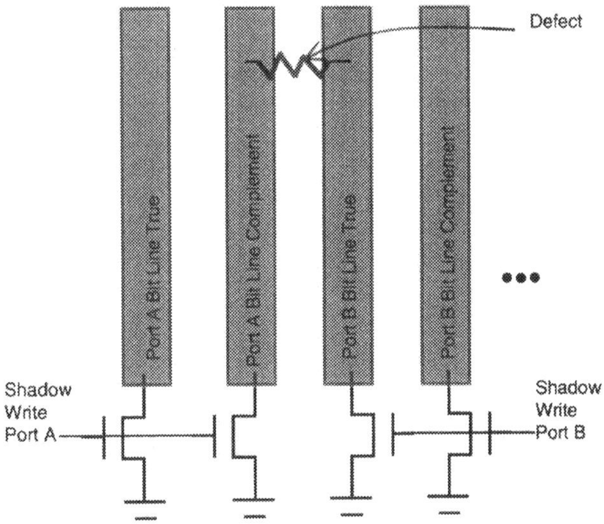

Figure 14-4. Shadow write grounding of alternating bit-line pairs.

5. GENERAL MEMORY DFT TECHNIQUES

There are a multitude of other DFT techniques that utilize slight modifications of existing memory circuitry or use memory features to facilitate testing in ways other than originally intended by the designer. For these reasons it is important to understand the memory design, to develop the best test methods to detect defects.

As already mentioned, the write word-line up time can be shortened to find gross resistive bit line contact problems. A shortened word-line timing can exacerbate other write margin issues, enabling the detection of defects in alternative locations. Any of the internal memory timings can be modified to accentuate problems. By moving the temperature and voltage slightly outside of the customer's specifications, robustness and quality can be enhanced by finding any marginal memories. Likewise, internal timings can be modified with DFT means to find other marginal product. DFT features can also allow for more accurate timings than could be measured by an external tester [202]. High-speed test clock multiplication can be applied through DFT techniques.

Internal voltage potentials can also be modified to look more effectively for defects. When a boosted word line is employed, the boost voltage can be modified [203]. Sense amplifiers can be margined by tweaking reference voltages during test [204]. Other on-chip generated voltages can be modified for DFT purposes as well [205]. These techniques are the staple of high quality memory testing, especially when more challenging memories are under consideration.

The use of a normal design function can also help find defects. If a memory has a half-select state, where the word line is high but the pre-charge circuitry remains on for unselected columns, subtle defects can be detected that are missed during full select reads. Since the pre-charge circuit fights against the memory cell, the bit lines appear more heavily loaded than just their normal capacitance would indicate. Other design functions can likewise be used to the test engineer's advantage.

The scan chain, which normally surrounds embedded memories, can be utilized to enhance test especially when trying to debug some obscure problem [206]. A sequence of ones and zeros can be scanned to the memory inputs and then clocked through the memory array. This scan chain can be quite helpful if the BIST is not as programmable and cannot provide the patterns that would be desired, in light of some peculiar characterization problem.

Certain memory and BIST features can be utilized to provide bit fail mapping support that is helpful for characterization, debug, and manufacturing line learning efforts. By sending the pass/fail BIST signal off chip, an address fail map can be assembled. Through sending some memory I/O off-chip during BIST test, a full bit fail map can be generated. As memory and chip design is being done, the bit fail mapping desires should be considered and the right DFT features included to help in manufacturability.

6. CONCLUSIONS

Modifying memory designs to facilitate test is normal and wise to ensure high quality. Certain defects which are missed during normal testing and are intermittent field fails can be easily detected when the right DFT feature is included and utilized for manufacturing test. Subtle defects can be found, thereby improving the quality. Design features which mask certain defects can be turned off during test to help find those defects. The design-for-test features truly provide high quality chips with the most advanced testing. It is important that effort be expended in the memory design phase to include the best possible DFT practices.

Chapter 15

Conclusions

Memory Self Test

"Memories light the corners of my mind ..." – from the song *The Way We Were*

1. THE RIGHT BIST FOR THE RIGHT DESIGN

Built-in self-test is a key enabler to high quality embedded memories. Given the incredible growth of embedded memories, BIST will only become more critical. Memories involve numerous analog circuits that store, retrieve, and compare data. These memories can have defects which affect their operation in many subtle ways.

To generate the correct built-in self-test a thorough procedure must be pursued. First the memory design, with its cell, pre-charge, sense amplifier, decoder, and write-driver circuitry must all be understood. These circuits need to be understood, not only in the way they are intended to operate but also in the ways that each of the circuits operate in the presence of defects. Comprehending the possible defect sites and the operational impact of those defects is no trivial matter. Ones and zeros are too high a level of abstraction to consider, when evaluating defects in memory circuitry. Instead, relative current flow and millivolts of potential difference need to be examined, based on the presence of defects.

These defective operations need to be abstracted into fault models that are usable in test. The fault models must reflect the defective circuit operation and must be used to determine proper testing. The proper testing, in turn, involves the application and observation of ones and zeros at the boundary of the memory.

Once memory design is understood and the proper fault models have been developed, the best test patterns can be generated. These test patterns take into account the various fault models for a given memory design. Since no single test pattern can suffice to detect all of the possible memory faults for a given design, a suite of test patterns needs to be employed. This pattern suite should be comprehensive in the defects it wants to detect.

After a suite of patterns is determined, other possible test capabilities should be considered. Certain contingencies should be developed in the event that new, unforeseen defects are found during manufacturing which drive new fault models. These contingencies require anticipating modifications to the planned test pattern.

This level of understanding of the design, the fault models, and the test patterns is required just to test a memory and ensure defective parts are not shipped. Redundancy goes beyond this point and is required to assure adequate yield on the vast majority of memories. The design, fault modeling, and patterns need to be revisited, in light of redundancy. Redundancy algorithms need to be developed to allocate each redundant dimension to the appropriate fails, thereby maximizing yield.

Once the design, fault models, test patterns, test contingencies, and redundancy algorithm are understood, the correct BIST can be designed. The test pattern suite can be implemented with a state machine or micro-code BIST. Which BIST is used depends on the maturity of the process technology and the complexity of the memory. The test contingencies need to be included in the BIST as programmable features, to be implemented as needed with manufacturing experience. The BIST also needs the proper redundancy algorithm built into it so that the best redundancy can be invoked on the fly, as each fail is detected. In this manner the best BIST for each memory can be designed and implemented, resulting in high quality memories with high yields.

2. MEMORY TESTING

Each type of memory design needs a slightly different test. As stated, different designs have different fault models. There are certain broad statements which can be made about each class of memory. These statements do not mean that there can be a one-size-fits-all approach to memory testing. Each specific memory design needs its own consideration. Even in straightforward single-port SRAMs, different design choices in the cell, sense amplifier, and decoder generate different fault models and therefore different tests.

More specifically, there are certain primary considerations for each memory type. For SRAMs, the differential bit-line pair, with its small signal swing, requires a write of one data type followed by a read of the opposite data type. Two back-to-back reads enable destructive read defects to be detected. These test are just part of an overall test pattern suite.

When using multi-port memories, the adjacencies between word lines and bit lines need careful understanding. The adjacent structures must be tested to ensure no shorting or defective interaction.

Silicon-on-insulator memories require tests to detect defects which are a function of the history and bipolar effects. These differences from bulk silicon memory require test pauses to allow the floating bodies to rise to their worst-case potentials, enabling subtle defect detection.

Content addressable memories have compare match and masking functions. These operations need to be tested logically while the CAM cells are tested for all of the defects that SRAMs see.

DRAMs have more subtle analog effects than any other memory type. These memories need to be tested for data retention, assuring the design specification, since even a defect-free cell has a certain amount of tolerable leakage.

Non-volatile memories have the broadest spread of tests for any category in this text. ROMs need a simple up-down counter addressing while feeding the outputs into a MISR. Other non-volatile memories today need a very limited number of test operations to prevent test from impacting the customer product life. Future non-volatile memories should be put through an extensive test pattern suite, based on experience with the ways a specific memory design fails due to manufacturing defects.

The memory testing discussions in this text have focused primarily on patterns and design-for-test techniques. In addition to these, the test environment is very key. The temperatures, voltages, and signal timings are all crucial to providing good quality memories for all possible customer applications. Between wafer and package tests, multiple temperatures and voltages need to be applied. No single worst-case set of conditions can be defined since various defects are more sensitive to different conditions. Elevated temperature and lowered voltage provide the slowest on-chip defect-free timings. Certain polysilicon defects are exacerbated, though, only at lower temperatures. For each process technology and memory type, the correct set of temperatures and voltages need to be determined. Normally, this can only be accomplished empirically so it is quite hard to make broad application statements with respect to environmental condition testing for memories.

3. THE FUTURE OF MEMORY TESTING

Small design changes are being implemented each year on existing memories. New memories are being developed on a regular basis. These small design changes and new memories each require more memory fault model development. Without the correct fault models, the designers and test engineers will falsely assume that a memory which passes test is defect free. As more bits are included on each new chip design, no naiveté can be tolerated with respect to memory design nor memory test. We need the best memory designs, the best fault models, the best tests, and the best BIST.

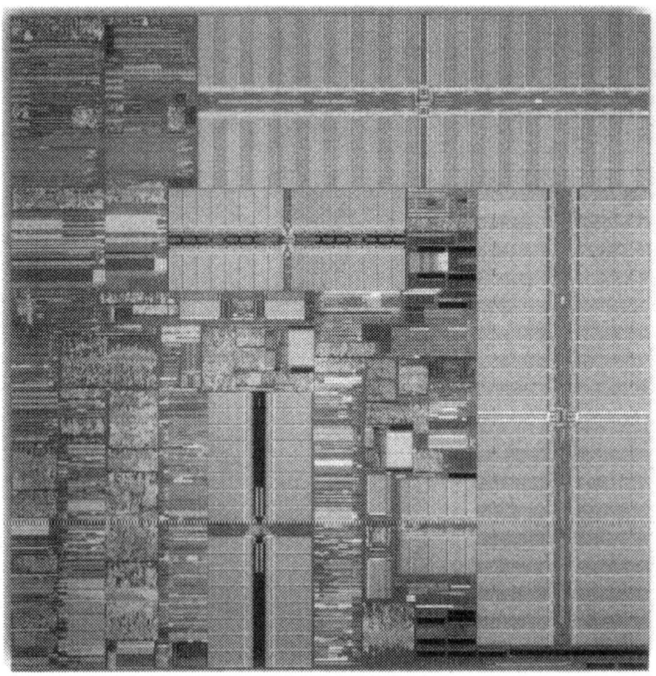

Figure 15-1. Photograph of PowerPC 750CX microprocessor with multiple memories and BIST.

Appendix A

Further Memory Fault Modeling

Extended Listing

The fault models in this appendix include those not described in chapter eight. It is recommended that the reader first become familiar with the faults covered in that chapter as they describe real manufacturing defects which must be covered during memory testing. The fault models covered here will help in the understanding of nomenclature and will aid readers as they review other literature on memory testing. Some of the fault models covered here are only mathematical curiosities while others provide helpful insight.

1. LINKED FAULTS

Since more than one defect can exist in a memory, the multiple faults can interact. As these interact they are referred to as linked fault models. Fault models that are linked can be of similar or dissimilar types [207]. They can also work in such a manner that one fault can mask the behavior of another fault. The occurrence of this kind of a fail must be rare in order to get reasonable chip yield. The probability of having two fails needs to be rarer still. The probability of having two fails, that in fact interact, must be exceedingly rare. Therefore linked faults should normally be of little concern. The only possibility for linked faults, which would be of concern, is if one defect activates two faults, which in turn are linked.

When multiple faults do not interact they are said to be *unlinked*. This is the normal case with multiple faults.

2. COUPLING FAULT MODELS

2.1 Inversion coupling fault

An inversion coupling fault involves an aggressor and a victim cell where the aggressor causes the victim cell's data to invert. A Markov diagram of this fault model is shown in figure A-1. This figure illustrates an example whereby any transition in cell "j" from a "0" to a "1" causes cell "i" to invert. No known defects will cause this type of defective operation nor is it known if such a fault has ever been seen.

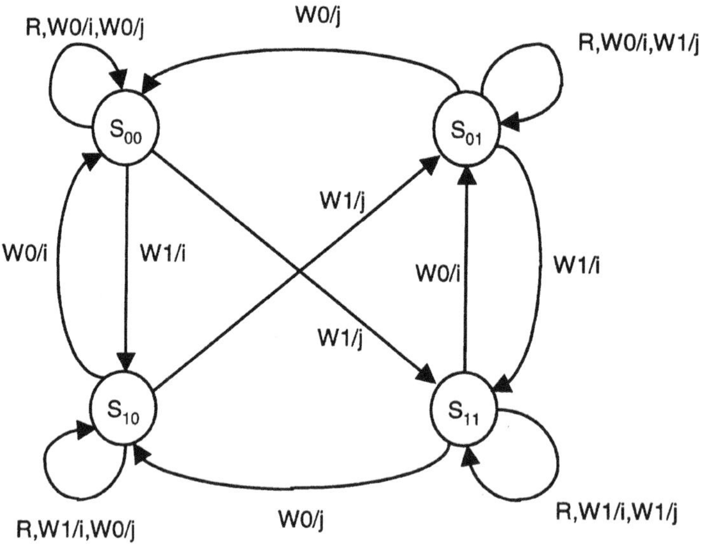

Figure A-1. Markov diagram of an inversion coupling fault.

2.2 Idempotent coupling fault

An idempotent coupling fault describes an event where writing a specific state in an aggressor cell causes a victim cell to go to a specific unintended state. Figure 8-8 shows an idempotent coupling fault where writing cell "j" to a "1" erroneously forces cell "i" to a "1" as well [208]. The source of the name for this fault is curious, although idempotent is a well understood mathematical term [209]. After some searching and dialog with others in this field, it seems that the reason for the name "idempotent" and its source have been lost in the annals of time. Since idempotent can refer to having the same singular result, even if an operation is repeated multiple times, it is possible that the idempotent coupling fault was so named in converse to the inversion coupling fault. The inversion coupling fault gets different results in the victim cell based on the number of times the aggressor cell is written. In contrast is the idempotent coupling fault, where a single erroneous value is maintained in the victim cell, as long as the victim cell itself is not purposely re-written.

2.3 Complex coupling fault

A complex coupling fault describes a defect where multiple cells must be in a specific state in order to sensitize a victim cell to a certain aggressor. A k-complex coupling fault is one where k is a value describing the number of cells which must interact [210]. If k is 5, then k-1 or 4 cells must be in a specific state in order for a 5th cell's transition to cause a failure in the victim cell.

2.4 State coupling fault

A state coupling fault does not require an operation to be performed nor a transition to occur. If an aggressor cell is in a specific state then the victim cell is forced to an erroneous state.

2.5 V coupling fault

A fault which requires two aggressor cells to transition, thereby forcing more charge into a victim cell and causing it to flip, is described by the V-type coupling fault model [211]. If only one of the aggressor cells transitions then the victim cell will not go to an erroneous state. Having two cells transition in direct adjacency to a victim cell can only happen in certain circumstances. If all of the bits in a word are written into a single sub array, then a cell in one row can be immediately adjacent to one cell and diagonally adjacent to another. If there is a two-to-one column decode, then a victim cell can be wedged between the two aggressor cells. Further, if a multi-port memory is employed then two ports can be writing to two cells adjacent to the victim cell and thus cause it to flip. In the case of a multi-port memory, the adjacencies can be from any pair of cells in the eight cells surrounding the base cell.

3. NEIGHBORHOOD PATTERN SENSITIVE FAULT MODELS EXPANDED

3.1 Pattern Sensitive Fault Model

The pattern sensitive fault model is the all encompassing form of the neighborhood pattern sensitive fault model (NPSF). When the neighborhood becomes all of the cells in the memory the fault model is referred to as the pattern sensitive fault model (PSF) and is sometimes called an unrestricted PSF [212, 213,214].

3.2 Active Neighborhood Pattern Sensitive Fault Model

The active or dynamic neighborhood pattern sensitive fault model [215, 216] refers to a defect that is exhibited when one of the neighborhood cells transitions causing a deleterious impact on the base or victim cell. The other cells in the neighborhood must be in specific states as is the case for a NPSF.

3.3 Passive Neighborhood Pattern Sensitive Fault Model

A passive neighborhood pattern sensitive fault model describes a defect where, as all of the neighborhood cells are in specific states, the base or victim cell is unable to transition. When active and passive neighborhood pattern sensitive faults are collectively discussed they sometimes are referred to by the acronym APNPSF.

3.4 Static Neighborhood Pattern Sensitive Fault Model

A static neighborhood pattern sensitive fault model describes a defect where, as all of the neighborhood cells are in specific states, the base or victim cell is forced to a specific erroneous state.

4. RECOVERY FAULT MODELS

4.1 Sense amplifier recovery fault model

A sense amplifier can saturate due to numerous reads of one data type [217]. When the opposite data type is read the sense amplifier has a defective tendency toward the previous data type and thus does not read correctly. This type of defective operation is described by the sense amplifier recovery fault model.

4.2 Write recovery fault model

When a write is performed, the address decoder can be slow to recover. This slowness can prevent a new address from being correctly read or written. The subsequent incorrect read or write is described as by the write recovery fault model.

4.3 Slow write recovery fault model

A slow write recovery fault model is similar to the previous write recover fault model [218]. The difference is that in a slow write recovery fault a subsequent read is correct but is delayed in time.

5. STUCK OPEN FAULT MODELS

5.1 Stuck open cell fault model

A cell can have a stuck off transfer device. The transfer device can be open or the gate can simply be stuck-at a zero [219]. In either case the result is a cell, which effectively has a stuck open connection to a bit line. In the case of an SRAM, since a true and a complement bit line are utilized the stuck-open defect may only affect one data type.

5.2 Stuck open bit line fault model

A column bit line can be open part way along its path to the cells [220,221] causing some of the cells can be connected to the bit line and others, beyond the break, are inaccessible. If the sense amplifier and the write driver circuitry are on the same side of the memory array then the cells up to the point of the break can function normally.

6. IMBALANCED BIT LINE FAULT MODEL

Defective cells along a DRAM bit line can be leaking onto the bit line and inducing an erroneous potential. This kind of a defect causes unintended cells to be "accessed" and they in turn drive the bit line into an imbalanced state resulting in incorrect reads. Thus, the name of an imbalanced bit line fault model is utilized. An SRAM can also be impacted by this type of fault if leakage occurs more on one of the bit lines in a pair than on the other.

7. MULTI-PORT MEMORY FAULTS

Faults, which can be sensitized by exercising a single port, are logically referred to as single-port faults [222] and are often use the notation "1PF". Faults, which require two ports to be exercised in order to sensitize a defect, are referred to as two-port faults. These are referred to by the notation "2PF". Faults can impact one or more cells. A single-port fault that impacts one cell or two cells is referred to by the notation "1PF1s" and "1PF2s", respectively. This notation is clearly extensible [223].

A strong fault is one that can be sensitized by performing a single port operation. A weak fault is one that creates insufficient disturbance to result in a fail when only one port is exercised. When multiple ports are exercised, multiple weak faults can cumulatively impact the operation and result in a memory failure. This is similar to the V-type coupling fault.

Appendix B

Further Memory Test Patterns

Extended Listing

The memory test patterns listed in this appendix are a supplement to those described in chapter nine of this text. The reader should first examine chapter nine and familiarize themselves with the patterns covered there. The patterns included in chapter nine are utterly essential to successfully producing high quality memories and thereby good chips. The patterns contained in this appendix are for reference and are especially helpful as one reads literature on memory testing. Some of the patterns discussed here are only mathematically interesting while others provide very helpful insight.

1. MATS PATTERNS

1.1 MATS

The modified algorithmic test sequence, also known as MATS, is a 4N pattern [224,225]. It is focused on finding stuck-at faults as well as detecting some address decoder faults. MATS has the same length as the Zero-One pattern but is far superior.

Table B-1. Description of the MATS pattern.

1	W0 ⇕
2	R0, W1 ⇕
3	R1 ⇕

1.2 MATS+

The MATS+ pattern requires 5N operations and is considered optimal for unlinked stuck-at faults [226].

Table B-2. Description of the MATS+ pattern.

1	W0 ⇕
2	R0, W1 ⇑
3	R1, W0 ⇓

1.3 MATS++

The MATS++ pattern is an improvement on the Marching 1/0 pattern (covered next in this appendix). It is a 6n pattern and eliminates certain redundancies [227]. (A *redundancy* is a repeat of an operation that does not allow any further faults to be detected. A pattern without redundancies is said to be *irredundant*.) This pattern detects some address decoder faults, stuck-at faults, and transition faults, along with some coupling faults.

Table B-3. Description of the MATS++ pattern.

1	W0 ⇕
2	R0, W1 ⇑
3	R1, W0, R0 ⇓

1.4 Marching 1/0

The Marching 1/0 pattern detects the same faults as the MATS++, but is longer. It is a 14n pattern.

Table B-4. Description of the Marching 1/0 pattern.

1	W0 ⇑
2	R0, W1, R1 ⇑
3	R1, W0, R0 ⇓
4	W1 ⇑
5	R1, W0, R0 ⇑
6	R0, W1, R1 ⇓

2. LETTERED MARCH PATTERNS

2.1 March A

The March A pattern is a 15n pattern. It focuses on detecting linked idempotent coupling faults. It also detects address decoder faults, stuck-at faults, and transition faults not linked with idempotent coupling faults.

Table B-5. Description of the March A pattern.

1	W0 ⇕
2	R0, W1, W0, W1 ⇑
3	R1, W0, W1 ⇑
4	R1, W0, W1, W0 ⇓
5	R0, W1, W0 ⇓

2.2 March B

The March B pattern can detect linked transition and idempotent coupling faults as well as detecting address decoder faults and stuck-at faults [228]. It is a 17n pattern.

Table B-6. Description of the March B pattern.

1	W0 ⇕
2	R0, W1, R1, W0, R0, W1 ⇑
3	R1, W0, W1 ⇑
4	R1, W0, W1, W0 ⇓
5	R0, W1, W0 ⇓

2.3 March C

The March C pattern [229] had the March C- pattern derived from it. The March C is not irredundant, as can be seen in the fourth march element. Chapter nine can be examined for a detailed discussion of the March C- pattern.

Table B-7. Description of the March C pattern.

1	W0 ⇕
2	R0, W1 ⇑
3	R1, W0 ⇑
4	R0 ⇕
5	R0, W1 ⇓
6	R1, W0 ⇓
7	R0 ⇕

2.4 March X

The March X pattern takes 6n cycles and is focused on finding unlinked inversion coupling faults.

Table B-8. Description of the March X pattern.

1	W0 ⇕
2	R0, W1 ⇑
3	R1, W0 ⇓
4	R0 ⇕

2.5 March Y

The March Y pattern enables testing of linked transition and inversion coupling faults. It also detects address decoder faults and stuck-at faults. It is an 8n pattern.

Table B-9. Description of the March Y pattern.

1	W0 ⇕
2	R0, W1, R1 ⇑
3	R1, W0, R0 ⇓
4	R0 ⇕

2.6 March C+, C++, A+, A++ Patterns

The March C+ and March C++ patterns are based on the March C pattern [230]. In the March C+, each read from the March C pattern is replaced with three reads to detect disconnected pull-up and pull-down paths inside a cell. The March C++ pattern includes two delay elements that look for retention type defects. Some refer to a different pattern that is called by the same name. This March C+ pattern is like the PMOVI pattern but with a read "0" element added at the end.

The March A+ and March A++ have the same changes, i.e. the triple reads and the added delay elements, as described in the preceding paragraph.

2.7 March LA

March LA is a 22n pattern with three consecutive writes in each of the key march elements [231]. It can detect all simple faults and many linked faults.

Table B-10. Description of the March LA pattern.

1	W0 ⇕
2	R0, W1, W0, W1, R1 ⇑
3	R1, W0, W1, W0, R0 ⇑
4	R0, W1, W0, W1, R1 ⇓
5	R1, W0, W1, W0, R0 ⇓
6	R0 ⇓

2.8 March SR+

The march test for simple realistic faults, also known as March SR+, is an 18n pattern [232]. It detects stuck-at faults, transition faults, coupling faults, and numerous other faults as well. Further, the double back-to-back read detects deceptive destructive reads.

Table B-11. Description of the March SR+ pattern.

1	W0 ⇓
2	R0, R0, W1, R1, R1, W0, R0 ⇑
3	R0 ⇓
4	W1 ⇑
5	R1, R1, W0, R0, R0, W1, R1 ⇓
6	R1 ⇑

This pattern was further enhanced by the inclusion of a delay to detect subtle retention faults. This is referred to as a March SRD+ pattern.

Table B-12. Description of the March SRD+ pattern.

1	W0 ⇓
2	R0, R0, W1, R1, R1, W0, R0 ⇑
3	Pause
4	R0 ⇓
5	W1 ⇑
6	R1, R1, W0, R0, R0, W1, R1 ⇓
7	Pause
8	R1 ⇑

3. IFA PATTERNS

3.1 9N Linear

Some patterns have become known simply by their numbers. One such pattern is the 9N linear test algorithm [233]. It is also often referred to as the inductive fault analysis-9 pattern or IFA-9. This pattern is commonly used and employs a pause for retention testing.

Table B-13. Description of the 9N linear test algorithm.

1	W0 ⇕
2	R0, W1 ⇑
3	R1, W0 ⇑
4	R0, W1 ⇓
5	R1, W0 ⇓
6	Pause
7	R0, W1 ⇑
8	Pause
9	R1 ⇑

3.2 13N

The 13N pattern detects coupling faults for bits within the same word [234,235]. This pattern is also referred to as the inductive fault analsysis-13 or IFA-13 pattern. Multiple background data types are required. Chapter nine can be examined for a discussion on background data types, which are needed in some patterns. The 13N pattern was developed to detect stuck-open cell errors.

Table B-14. Description of the 13N pattern.

1	W0 ⇕
2	R0, W1, R1 ⇑
3	R1, W0, R0 ⇑
4	R0, W1, R1 ⇓
5	R1, W0, R0 ⇓
6	Pause
7	R0, W1 ⇑
8	Pause
9	R1 ⇑

4. OTHER PATTERNS

4.1 MovC

The movC was developed to ease BIST pattern implementation [236]. It is not, however, irredundant. The pattern requires 33N for each data background type.

Table B-15. Description of the movC pattern.

1	W0 ⇕
2	R0, W0, W1, R1 ⇓
3	R1, W1, W0, R0 ⇓
4	R0, W0, W1, R1 ⇓
5	R1, W1, W0, R0 ⇓
6	R0, W0, W1, R1 ⇑
7	R1, W1, W0, R0 ⇑
8	R0, W0, W1, R1 ⇑
9	R1, W1, W0, R0 ⇑

4.2 Moving Inversion

The moving inversion or MOVI pattern [237] was the precursor to the PMOVI pattern discussed in chapter nine. It involved all of the elements of the PMOVI pattern but repeated them based on the number of address inputs to the memory.

4.3 Butterfly

The butterfly pattern is quite complicated but does take fewer steps than the galloping pattern. The essence of the butterfly pattern is that the base cell is modified following a walking algorithm. After each write of a base cell, the four cells adjacent to it are read. These are in the north, south, east, and west directions. Once these four cells are read the base cell is again read. The butterfly pattern may be continued by reading the next farther out cells in these four directions, followed again by the base cell. The distance continues to be doubled until the edge of the memory or sub-array is reached. It can be seen that this pattern is rather convoluted and only DRAMs have been helped uniquely through this test.

5. SMARCH

The SMARCH pattern was described in chapter nine. The pseudo-code for this pattern is described below.

Table B-16. Pseudo-code description of the SMARCH pattern.

```
// Initialize memory to all 0s
for i = 0..N {
   for j = 0..m { Rx, W0 }
   for j = 0..m { W0, R0 }
}
// like a R0, W1, R1, W1 ⇑ march pattern element
for i = 0..N {
   for j = 0..m { R0, W1 }
   for j = 0..m { R1, W1 }
}
// like a R1, W0, R0, W0 ⇑ march pattern element
for i = 0..N {
   for j = 0..m { R1, W0 }
   for j = 0..m { R0, W0 }
}
// like a R0, W1, R1, W1 ⇓ march pattern element
for i = N..0 {
   for j = 0..m { R0, W1 }
   for j = 0..m { R1, W1 }
}
// like a R1, W0, R0, W0 ⇓ march pattern element
for i = N..0 {
   for j = 0..m { R1, W0 }
   for j = 0..m { R0, W0 }
}
// Read all the cells at 0
for i = 0..N {
   for j = 0..m { R0, W0 }
   for j = 0..m { W0, R0 }
}
```

6. PSEUDO-RANDOM

Pseudo-random patterns were described in chapter nine. In table B-16 a five-bit pseudo-random sequence is provided for reference. Note that the first and 32nd entries match as do the second and 33rd entries. After 31 cycles the pseudo-random sequence has re-started. Note also that the all zeros state is not present.

Table B-17. Example five-bit pseudo-random sequence.

Pass	X[0]	X[1]	X[2]	X[3]	X[4]
1	1	0	0	0	0
2	0	1	0	0	0
3	1	0	1	0	0
4	0	1	0	1	0
5	1	0	1	0	1
6	1	1	0	1	0
7	1	1	1	0	1
8	0	1	1	1	0
9	1	0	1	1	1
10	1	1	0	1	1
11	0	1	1	0	1
12	0	0	1	1	0
13	0	0	0	1	1
14	1	0	0	0	1
15	1	1	0	0	0
16	1	1	1	0	0
17	1	1	1	1	0
18	1	1	1	1	1
19	0	1	1	1	1
20	0	0	1	1	1
21	1	0	0	1	1
22	1	1	0	0	1
23	0	1	1	0	0
24	1	0	1	1	0
25	0	1	0	1	1
26	0	0	1	0	1
27	1	0	0	1	0
28	0	1	0	0	1
29	0	0	1	0	0
30	0	0	0	1	0
31	0	0	0	0	1
32	1	0	0	0	0
33	0	1	0	0	0

Appendix C

State Machine HDL

Example Code

In chapter 11 a discussion was included of state machine BISTs. A series of interrelated counters define the stimulus to the memory. Table 11-2 included some VHDL code for a three-bit ripple counter. The complete VHDL follows in figure C-1.

```
library ieee, work;
USE IEEE.Std_Logic_1164.all ;

entity threebit_e is
// Code by Thomas J Eckenrode
  port ( STclk      :in  std_ulogic;
        inc        :in  std_ulogic;
        countout   :out std_ulogic_vector(2 downto 0));
end entity threebit_e;
architecture threebit_a of threebit_e is
signal u_bit     : std_ulogic_vector(2 downto 0);
signal l_bit     : std_ulogic_vector(2 downto 0);

begin
   latchdef:process(STclk, u_bit)
    begin
     if STclk='1' AND STclk'EVENT then
         l_bit  <= u_bit;
     end if;
   end process latchdef;
   with inc select
         u_bit(0) <=  not l_bit(0) when '1',
```

```
                    l_bit(0) when others;
    with inc select
        u_bit(1) <= (l_bit(0) XOR l_bit(1)) when '1',
                  l_bit(1) when others;
    with inc select
        u_bit(2) <= ((l_bit(0) AND l_bit(1)) XOR l_bit(2)) when '1',
                  l_bit(2) when others;
    countout <= l_bit;

                        end architecture threebit_a;
```

Figure C-1. VHDL deck for a three-bit ripple counter.

A key portion of a memory BIST is the read/write controller. It generates the read-enable and write-enable signals. A segment of code was included in Table 11-4. The complete Verilog code for generating the read-enable, write-enable, and data signals for the first two elements of the March C- pattern are given in figure C-2.

```
module RWGEN( clock, start, max_addr, read, write, data );
// Code by Garret S. Koch
input  clock, start, max_addr;
output read, write, data;

reg  next_state0,
     next_state1,
     next_state2,
     next_state3;

reg  state0,
     state1,
     state2,
     state3;

initial
 begin
   state0 = 1 ;
   state1 = 0 ;
   state2 = 0 ;
   state3 = 0 ;
 end
```

```
    always @ (posedge clock)
     begin
      state0 = next_state0 ;
      state1 = next_state1 ;
      state2 = next_state2 ;
      state3 = next_state3 ;
     end

    always @ (start or state0 or state3 or max_addr)
       if    ( start   )          next_state0 = 0 ;
       else if ( max_addr & state3 ) next_state0 = 1 ;
       else                       next_state0 = state0;

    always @ (start or state1 or max_addr)
       if    ( start   )          next_state1 = 1 ;
       else if ( max_addr )          next_state1 = 0 ;
       else                       next_state1 = state1;

    always @ (state1 or state3 or max_addr)
       if    ( max_addr & state1 ) next_state2 = 1 ;
       else if (!max_addr & state3 ) next_state2 = 1 ;
       else                       next_state2 = 0 ;

    always @ (state2 or max_addr)
       if    (          state2 ) next_state3 = 1 ;
       else                       next_state3 = 0 ;

    assign write = state1  | state3;
    assign read  = state2;
    assign data  = !(state1 | state2);

endmodule
```

Figure C-2. Sample Verilog for read/write generator.

The BIST address counter should be able to increment or decrement addresses and perform these operations on either a ripple word or ripple bit basis. When decrementing, the maximum address must first be loaded into the counter. All of these features are included in the counter shown in Figure C-3 [238].

```
module address_counter (a, b, c, f, inc_dec, increment, word_limit,
   bit_limit, w_limit, b_limit, word_address, bit_address,
   next_word_r, next_bit_r, scan_in, scan_out,add_comp, word_match);

// Code by R. Dean Adams, Robert M. Mazzarese, Esq.

   input a, b, c, f, scan_in, inc_dec, add_comp, word_match, increment;
   input [7:0] word_limit, bit_limit, w_limit, b_limit;

   output [7:0] word_address, bit_address, next_word_r, next_bit_r;
   output scan_out ;

   reg [7:0] next_word_r, next_bit_r, word_address, bit_address;
   reg [7:0] tempword, tempbit;

   initial begin
      next_word_r = 8'b00000000; next_bit_r = 8'b00000000;
      word_address = 8'b00000000; bit_address = 8'b00000000;
   end

   //TRANSFER pipeline values to output

   always @ (posedge f)  begin
      #1;
      word_address <= next_word_r;
      bit_address <= next_bit_r;
   end

//CALCULATE new address when increment signal received
   always @ (posedge increment) begin #1;
      casez ({add_comp,inc_dec,word_match})
         3'b10z: begin
            next_word_r = 8'b00000000;
            next_bit_r = 8'b00000000; end
         3'b11z: begin
            next_word_r = w_limit;
            next_bit_r = b_limit; end
         3'b000: begin
            next_word_r = next_word_r+1; end
         3'b010: begin
            next_word_r = next_word_r - 1; end
```

```
            3'b001: begin
                next_bit_r = next_bit_r + 1;
                next_word_r = 8'b00000000; end
            3'b011: begin
                next_bit_r = next_bit_r - 1;
                next_word_r = w_limit; end
            default: begin
                $display("address counter default");
                next_word_r = 8'b00000000;
                next_bit_r = 8'b00000000; end
        endcase
    end

/* RELOAD COUNTER whenever inc_dec changes */
    always @(inc_dec) begin #1;
        case (inc_dec)
            1'b1: begin
                next_word_r = 0; next_bit_r = 0;
                next_bit_r <= bit_limit; next_word_r <= word_limit;
                end
            1'b0: begin
                next_word_r = 1; next_bit_r = 1;
                next_bit_r <= 8'b00000000; next_word_r <= 8'b00000000;
                end
        endcase
    end
endmodule
```

Figure C-3. Address counter example.

References

1 G.E. Moore, "Cramming more components onto integrated circuits," Electronics, Vol. 38 No. 8, 4/19/65.

2 G. Bell, "Some thoughts by Gordon Bell," DOE Clusters Workshop, http://www.scl.ameslab.gov/workshops/Talks/Bell/sld001.htm, 4/10/97.

3 G.E. Moore, "The continuing Silicon technology evolution inside the PC platform," http://developer.intel.com/update/archive/issue2/feature.htm.

4 P.P. Gelsinger et al., "Microprocessors circa 2000," IEEE Spectrum, 10/89, pp. 43-7.

5 G. Vandling, "Modeling and testing the Gekko microprocessor, an IBM PowerPC derivative for Nintendo," International Test Conference 2001, pp. 593-9.

6 K. Thompson, International Test Conference 1995 Keynote, Electronic Engineering Times, 10/30/95, p. 1.

7 P. Gelsinger, "The challenges of design and test for the world wide web," International Test Conference 2001 Keynote, p. 12.

8 S. Parihar, et al., "A high density 0.10um CMOS technology using low K dielectric and copper interconnect," International Electron Devices Meeting 2001, pp. 249-52.

9 International Technology Roadmap for Semiconductors, 2001 Edition, http://public.itrs.net/Files/2001ITRS/Home.htm.

10 K. Osada, "Universal-Vdd 0.65-2.0 V 32-kB cache using a voltage-adapted timing-generation scheme and a lithographically symmetrical cell," IEEE Journal of Solid-State Circuits Vol. 36, No. 11, 11/2001, pp. 1738-43.

11 K. Furumochi et al., "A 500 M Hz 288kb CMOS SRAM macro for on-chip cache," International Solid State Circuits Conference 1996, pp. 156-7.

12 H. Pilo, R.D. Adams et al., "Bitline Contacts in High-Density SRAMS: Design for Testability and Stressability," International Test Conference 2001, pp. 776-82.

13 A.J. Bhavnagarwala, X. Tang, J.D. Meindl, "The impact of intrinsic device fluctuations on CMOS SRAM cell stability," IEEE Journal of Solid-State Circuits, Vol. 36, No. 4, 4/2001, pp. 658-65.

14 M. Tamjidi, personal communication August 2000.

15 D. Draper et al., "Circuit techniques in a 266-MHz MMX-Enabled processor," IEEE Journal of Solid-State Circuits, Vol. 32, No. 11, 11/1997, pp. 1650-64.

[16] N. Shibata, M. Watanabe, Y. Tanabe, "A current-sensed high-speed and low-power first-in-first-out memory using a wordline/bitline-swapped dual-port SRAM cell," IEEE Journal of Solid-State Circuits, Vol. 37, No. 6, 6/2002, pp. 735-50.

[17] K. Yokomizo, K. Naito, "Design techniques for high-throughput BiCMOS self-timed SRAM's," IEICE Trans. Electron, Vol. E76-C, No. 5, 5/1993, pp. 824-9.

[18] K.J. Schultz, P.G. Gulak, "Fully-parallel multi-megabit integrated CAM/RAM design," IEEE Int. Workshop on Memory Technology, 8/1994, pp. 46-51.

[19] N. Shibata et al., "A 2-V 300-MHz 1-Mb current-sensed double-density SRAM for low-power 0.3 um CMOS/SIMOX ASICs," IEEE Journal of Solid-State Circuits, Vol. 36, No. 10, 10/2001, pp. 1524-37.

[20] N. Shibata et al., "A 2-V 300-MHz 1-Mb current-sensed double-density SRAM for low-power 0.3 um CMOS/SIMOX ASICs," IEEE Journal of Solid-State Circuits, Vol. 36, No. 10, pp. 1524-37.

[21] K. Sasaki, "A 7 nanosecond 140 milliwatt 1 Megabit CMOS SRAM with current sense amplifier," International Solid State Circuits Conference 1992, pp. 208-209.

[22] R.D. Adams, E.S. Cooley, P.R. Hansen, "A self-test circuit for evaluating memory sense-amplifier signal," International Test Conference 1997, pp. 217-25.

[23] B.S. Amutur, M.A. Horowitz, "Fast low-power decoders for RAMs," IEEE Journal of Solid-State Circuits, Vol. 36, No. 10, 10/2001, pp. 1506-15.

[24] R.D. Adams et al., "An 11-ns 8k x 18 CMOS static RAM," International Solid State Circuits Conference 1988, pp. 242-3.

[25] H. Pilo et al., "A 300 M Hz, 3.3V 1Mb SRAM," International Solid State Circuits Conference 1996, pp. 148-9.

[26] R. Mookerjee, "Segmentation: a technique for adapting high-performance logic ATE to test high-density, high-speed SRAMs," IEEE Workshop on Memory Test 1993, pp. 120-4.

[27] K. Noda et al., "An ultrahigh-density high-speed loadless four-transistor SRAM macro with twisted bitline architecture and triple-well shield," IEEE Journal of Solid-State Circuits, Vol. 36, No. 3, 3/2001, pp. 510-5.

[28] M. Redeker, B.F. Cockburn, D.G. Elliott, "An investigation into crosstalk noise in DRAM structures," IEEE Memory Technology, Design, and Test Workshop 2002, pp. 123-9.

[29] R. Flaker, IBM senior technical staff member, 1995 personal communication.

[30] IEEE Memory Technology, Design, & Test Workshop 2001, Panel on Memory Redundancy & Repair Challenges.

[31] R.D. Adams et al., "A 5 nanosecond Store Barrier Cache with Dynamic Prediction of Load / Store Conflicts in Superscalar Processors," IEEE International Solid-State Circuits Conference 1997, pp. 414-5.

[32] J.L. Hennessy, D.A. Patterson, *Computer Architecture: A Quantitative Approach*, Morgan Kaufmann, 1996.

[33] T.P. Haraszti, *CMOS Memory Circuits*, Kluwer, 2000.

[34] P. Lin, J. Kuo, "A 1-V 128-kv four-way set-associative CMOS cache memory using wordline-oriented tag-compare (WLOTC) structure with the content-addressable-memory (CAM) 10-transistor tag cell," IEEE Journal of Solid-State Circuits, Vol. 36, No. 4, 4/2001, pp. 666-75.

[35] R. Gibbins, R.D. Adams et al., "Design and Test of a 9-port SRAM for a 100Gb/s STS-1 Switch," IEEE Memory Technology, Design, and Test Workshop 2002, pp. 83-7.

[36] S. Wood, et al., "A 5Gb/s 9-port application specific SRAM with built-in self-test," IEEE Memory Technology, Design, and Test Workshop 1995, pp. 68-73.

[37] M. Canada et al., "A 580 MHz RISC microprocessor in SOI," IEEE International Solid-State Circuits Conference 1999, pp. 430-1.

[38] S.K. Mathew et al., "Sub-500-ps 64-b ALUs in 0.18-um SOI/Bulk CMOS: Design and scaling trends," IEEE Journal of Solid-State Circuits, Vol. 36, No. 11, 11/2001, pp. 1636-46.

[39] K. Bernstein, N. Rohrer, *SOI Circuit Design Concepts*, Kluwer, 2000.

[40] J. Colinge, *Silicon-On-Insulator Technology: Materials to VLSI*, Kluwer, 1997.

[41] D.A. Johns, K. Martin, *Analog Integrated Circuit Design*, Wiley, 1997.

[42] M.S.L. Lee et al., "A physically based compact model of partially depleted SOI MOSFETs for analog circuit simulation," IEEE Journal of Solid-State Circuits, Vol. 36, No. 1, 1/2001, pp. 110-21.

[43] M. Wood, G. Smith, J. Pennings, "Converting a SRAM from bulk Si to partially depleted SOI," IEEE Custom Integrated Circuits Conference 1999, pp. 227-230.

[44] T. Ohsawa et al., "Memory design using one-transistor gain cell on SOI," International Solid State Circuits Conference 2002, pp. 152-3.

[45] R.D. Adams, P. Shephard III, "Silicon on insulator technology impacts on SRAM testing," VLSI Test Symposium 2000, pp. 43-47.

[46] R.J. Sung, et al. "Design of an embedded fully-depleted SOI SRAM," IEEE Memory Technology, Design, and Test Workshop 2001, pp. 13-8.

[47] A.G. Aipperspach et al., "A 0.2um, 1.8-V, SOI, 550-MHz, 64-b PowerPC microprocessor with copper interconnects," IEEE Journal of Solid-State Circuits, Vol. 34, No. 11, 11/1999, pp. 1430-5.

[48] A. Marshall, S. Natarajan, *SOI Design: Analog, Memory, and Digital Techniques*, Kluwer, 2002.

[49] E. MacDonald, N.A. Touba, "Delay testing of SOI circuits: Challenges with the history effect," International Test Conference 1999, pp. 269-75.

[50] E. MacDonald, "Testability and fault modeling of partially depleted silicon-on-insulator integrated circuits," Ph.D. Dissertation, University of Texas at Austin, 5/2002.

[51] F. Shafai et al., "Fully parallel 30-MHz, 2.5-Mb CAM," IEEE Journal of Solid-State Circuits, Vol. 33, No. 11, 11/1999, pp. 1690-6.

[52] J. Handy, *The Cache Memory Book*, Academic Press, 1993.

[53] J. Podaima, G. Gulak, "A self-timed, fully-parallel content addressable queue for switching applications," IEEE Custom Integrated Circuits Conference 1999, pp. 239-42.

[54] D. Bradley, P. Mahoney, B. Stackhouse, "The 16kB single-cycle read access cache on a next-generation 64b Itanium microprocessor," International Solid State Circuits Conference 2002, pp. 110-1.

[55] H. Miyatake, M. Tanaka, Y. Mori, "A design for high-sped low-power CMOS fully parallel content-addressable memory macros," IEEE Journal of Solid-State Circuits, Vol. 36, No. 6, 6/2001, pp. 956-68.

[56] V. Lines et al., "66MHz 2.3M ternary dynamic content addressable memory," IEEE Memory Technology, Design, and Test Workshop 2000, pp. 101-5.

[57] L. Ternullo, R.D. Adams et al., "Deterministic self-test of a high-speed embedded memory and logic processor subsystem," International Test Conference 1995, pp. 33-44.

[58] International Technology Roadmap for Semiconductors, 2001 Edition, http://public.itrs.net/Files/2001ITRS/Home.htm.

[59] H. Yoon, et al., "A 2.5-V, 333-Mb/s/pin, 1-Gbit, double-data-rate synchronous DRAM," IEEE Journal of Solid-State Circuits, Vol. 34, No. 11, 11/1999, pp. 1589-99.

[60] B. Prince, *High Performance Memories*, Wiley, 1999.

61 B. Keeth, R.J. Baker, *DRAM Circuit Design: A Tutorial*, IEEE Solid-State Circuit Society, 2001.

62 M. Redeker, B. Cockburn, et al., "Fault modeling and pattern-sensitivity testing for a multilevel DRAM," IEEE Memory Technology, Design, and Test Workshop 2002, pp. 117-22.

63 T. Sekiguchi, et al., "A low-impedance open-bitline array for multigigabit DRAM," IEEE Journal of Solid-State Circuits, Vol. 37, No. 4, 4/2002, pp. 487-98.

64 D. Takashima, H. Nakano, "A cell transistor scalable DRMA array architecture," IEEE Journal of Solid-State Circuits, Vol. 37, No. 5, 5/2002, pp. 587-91.

65 H. Hoenigschmid, et al., "A $7F^2$ cell and bitline architecture featuring tilted array devices and penalty-free vertical BL twists for 4-Gb DRAMs," IEEE Journal of Solid-State Circuits, Vol. 35, No. 5, 5/2000, pp. 713-18.

66 T. Takahashi, et al., "A multigigabit DRAM technology with $6F^2$ open-bitline cell, distributed overdriven sensing and stacked-flash fuse," IEEE Journal of Solid-State Circuits, Vol. 36, No. 11, pp. 1721-7.

67 C. Chang, J. Wang, C. Yang, "Low-power and high-speed ROM modules for ASIC applications," IEEE Journal of Solid-State Circuits, Vol. 36, No. 10, 10/2001, pp. 1516-23.

68 B. Yand, L. Kim, "A low-power ROM using charge recycling and charge sharing," International Solid State Circuits Conference 2001, pp. 108-9.

69 T.P. Haraszti, *CMOS Memory Circuits*, Kluwer, 2000.

70 W.D. Brown, J.E. Brewer, Nonvolatile Semiconductor Memory Technology, IEEE Press, 1997.

71 T. Jung, et al., "A 3.3-V single power supply 16-Mb nonvolatile virtual DRAM using a NAND flash memory technology," IEEE Journal of Solid-State Circuits, Vol. 32, No. 11, 11/1997, pp. 1748-57.

72 J. Lee, et al., "A 1.8V 1Gb NAND flash memory with 0.12um STI process technology," International Solid State Circuits Conference 2002, pp. 104-5.

73 K. Takeuchi, T. Tanaka, "A dual-page programming scheme for high-speed multigigabit-scale NAND flash memories," IEEE Journal of Solid-State Circuits, Vol. 36, No. 5, 5/2001, pp. 744-51.

74 S. Atsumi, et al., "A channel-erasing 1.8-V-only 32-Mb NOR flash EEPROM with a bitline direct sensing scheme," IEEE Journal of Solid-State Circuits, Vol. 35, No. 11, 11/2000, pp. 1648-54.

75 P. Cappelletti, et al., *Flash Memories*, Kluwer, 1999.

76 J. Tsouhlarakis, et al., "A flash memory technology with quasi-virtual ground array for low-cost embedded applications," IEEE Journal of Solid-State Circuits, Vol. 36, No. 6, 6/2001, pp. 969-78.

77 G. Campardo, et al., "40-mm2 3-V-only 50-MHz 64-Mb 2-b/cell CHE NOR flash memory," IEEE Journal of Solid-State Circuits, Vol. 35, No. 11, 11/2000, pp. 1655-67.

78 M. Borgatti, et al., "A 64-min single-chip voice recorder/player using embedded 4-b/cell flash memory," IEEE Journal of Solid-State Circuits, Vol. 36, No. 3, 3/2001, pp. 516-21.

79 J.M. Portal, et al., "An automated design methodology for EEPROM cell (ADE)," IEEE Memory Technology, Design, and Test Workshop 2002, pp. 137-42.

80 M. Mohammad, K.K. Saluja, "Flash memory disturbances: modeling and test," VLSI Test Symposium 2001, pp. 218-24.

81 A. Sheikholeslami, "Ferroelectric memory design," International Solid State Circuits Conference Tutorial, 2/3/2002.

82 B. Prince, *Emerging Memories: Technology and Trends*, Kluwer, 2002.

[83] S. Kawashima, et al., "Bitline GND sensing technique for low-voltage operation FeRAM," IEEE Journal of Solid-State Circuits, Vol. 37, No. 5, 5/2002, pp. 592-8.

[84] D. Takashima, et al., "A sub-40-ns chain FRAM architecture with 7-ns cell-plate-line drive," IEEE Journal of Solid-State Circuits, Vol. 34, No. 11, 11/1999, pp. 1557-63.

[85] M. Choi, et al., "A 0.25um 3.0V 1T1C 32Mb nonvolatile ferroelectric RAM with address transition detector(ATD) and current forcing latch sense amplifier(CFKSA) Scheme," International Solid State Circuits Conference 2002, pp. 162-3.

[86] B.F. Cockburn, "Advanced embedded memory technologies tutorial," IEEE Memory Technology, Design, and Test Workshop 2002.

[87] J. Kang, et al., "A hierarchy bitline boost scheme for sub-1.5V operation and short precharge time on high density FeRAM," International Solid State Circuits Conference 2002, pp. 158-9.

[88] G. Braun, et al., "A robust $8F^2$ ferroelectric RAM cell with depletion device(DeFeRAM)," IEEE Journal of Solid-State Circuits, Vol. 35, No. 5, 5/2000, pp. 691-6.

[89] B. Jeon, et al., "A 0.4-um 3.3V 1T1C 4-Mb nonvolatile ferroelectric RAM with fixed bitline reference voltage scheme and data protections circuit," IEEE Journal of Solid-State Circuits, Vol. 35, No. 11, 11/2000, pp. 1690-4.

[90] M. Clendenin, "Common memories turn exotic," Electronic Engineering Times, 4/29/2002, p. 60.

[91] S. Tehrani, et al., "Progress and outlook for MRAM technology," IEEE Transactions on Magnetics, Vol. 35, No. 5, 9/1999, pp. 2814-9.

[92] R.A. Sinclair, et al., "A practical 256K GMR NV memory for high shock applications," International Non-Volatile Memory Technology Conference 1998, pp. 38-42.

[93] R. Zhang, W.C. Black, M.W. Hassoun, "Windowed MRAM sensing scheme," IEEE Memory Technology, Design, and Test Workshop 2000, pp. 47-52.

[94] C. Brown, "MTJ memory seen as possible DRAM, SRAM replacement," Electronic Engineering Times, 7/23/2001, p. 65.

[95] R. Scheuerlein, et al., "A 10ns read and write non-volatile memory array using a magnetic tunnel junction and FET switch in each cell," International Solid State Circuits Conference 2000, pp. 128-9.

[96] P.K. Naji, et al., "A 256kb 3.0V 1T1MTJ nonvolatile magnetoresistive RAM," International Solid State Circuits Conference 2001, pp. 122-3.

[97] S. Lai, T. Lowrey, "OUM – a 180 nm nonvolatile memory cell element technology for stand alone and embedded applications," International Electron Devices Meeting 2001, pp. 803-6.

[98] M. Gill, T. Lowrey, J. Park, "Ovonic unified memory – a high-performance nonvolatile memory technology for stand-alone memory and embedded applications," International Solid State Circuits Conference 2002, pp. 202-3.

[99] M. Brehob, et al., "The potential of carbon-based memory systems," IEEE Memory Technology, Design, and Test Workshop 1999, pp. 110-4.

[100] R. McConnell, U. Moller, D. Richter, "How we test Siemens' embedded DRAM cores," International Test Conference 1998, pp. 1120-5.

[101] R.L. Geiger, P.E. Allen, N.R. Strader, *VLSI Design techniques for analog and digital circuits*, McGraw-Hill, 1990.

[102] A.J. van de Goor, *Testing Semiconductor Memories: Theory and Practice*, ComTex Publishing, Gouda, The Netherlands, 1998.

[103] John Barth, personal communication April 3, 2002.

[104] Bryan Robbins, personal communication 4Q 1997.

[105] R. David, A. Fuentes, B. Courtois, "Random pattern testing versus deterministic testing of RAMs," IEEE Transactions on Computers, Vol. 38, No. 5, May 1989, pp. 637-50.

[106] R. Rajsuman, "Algorithms to test PSF and coupling faults in random access memories," IEEE Int. Workshop on Memory Testing 1993, pp. 49-54.

[107] M. Franklin, K.K. Saluja, K. Kinoshita, "Design of a RAM with row/column pattern sensitive fault detection capability," International Test Conference 1989, pp. 327-36.

[108] S. Murray, "A user's approach to characterization and test of commercially available SRAMs," IEEE Memory Technology, Design, and Test Workshop 1998, p. 68.

[109] R.D. Adams, E.S. Cooley, "Analysis of a deceptive destructive read memory fault model and recommended testing," IEEE North Atlantic Test Workshop 1996.

[110] R.D. Adams, E.S. Cooley, "False write through and un-restored write electrical level fault models for SRAMs," IEEE Memory Design, Technology, and Test Workshop 1997, pp. 27-32.

[111] D. Niggemeyer, M. Redeker, J. Otterstedt, "Integration of non-classical faults in standard march tests," International Test Conference 1998, pp. 91-6.

[112] Jon Lachman, personal communication August 6, 2001.

[113] S. Hamdioui, A.J. van de Goor, "Address decoder faults and their tests for two-port memories," IEEE Memory Technology, Design, and Test Workshop 1998, pp. 97-103.

[114] S. Al-Harbi, S.K. Gupta, "An efficient methodology for generating optimal and uniform march tests," VLSI Test Symposium 2001, pp. 231-7.

[115] J. Otterstedt, D. Niggemeyer, T. Williams, "Detection of CMOS address decoder open faults with march and pseudo random memory tests," International Test Conference 1998, pp. 53-62.

[116] M. Sachdev, "Test and testability techniques for open defects in RAM address decoders," International Test Conference 1996, pp. 428-34.

[117] R.D. Adams, E.S. Cooley, "The limits of digital testing for dynamic circuits," VLSI Test Symposium 1999, pp. 28-32.

[118] P. Nagaraj, et al. "Defect analysis and a new fault model for multi-port SRAMs," IEEE International Symposium on Defect and Fault Tolerance in VLSI Systems 2001, pp. 366-74.

[119] F. Karimi, et al., "A parallel approach for testing multi-port static random access memories," IEEE Memory Technology, Design, and Test Workshop 2001, pp. 73-81.

[120] J. Zhao, et al., "Detection of inter-port faults in multi-port static RAMs," VLSI Test Symposium 2000, pp. 297-302.

[121] C.A. Dean, Y. Zorian, "Do you practice safe test? What we found out about your habits," International Test Conference 1994, pp. 887-92.

[122] H. Bonges III, R.D. Adams, et al., "A 576K 3.5ns access BiCMOS ECL static RAM with array built-in self-test," IEEE Journal of Solid-State Circuits, Vol. 27, No. 4, 4/1992, pp. 649-56.

[123] A.J. van de Goor, *Testing Semiconductor Memories: Theory and Practice*, ComTex Publishing, Gouda, The Netherlands, 1998.

[124] H. Maeno, S. Iwade, S. Kayano, "Embedded RAM test using flag-scan register," Electronics and Communications in Japan, Part 3, Vol. 77, No. 4, 1994, pp. 110-18.

[125] R. Treuer, V.K. Agarwal, "Built-in self-diagnosis for repairable embedded RAMs," IEEE Design & Test of Computers, 6/1993, pp. 24-32.

[126] Y. Matsuda, et al., "A new array architecture for parallel testing in VLSI memories," International Test Conference 1989, pp. 322-6.

[127] L.Ternullo, R.D. Adams, et al., "Deterministic self-test of a high-speed embedded memory and logic processor subsystem," International Test Conference 1995, pp. 33-44.

[128] A.J. van de Goor, et al., "March LR: A test for realistic linked faults," VLSI Test Symposium 1996, pp. 272-80.

[129] A.J. van de Goor, "Using march tests to test SRAMs," IEEE Design & Test of Computers, 3/1993, pp. 8-13.

[130] J. Sosnowski, "In system testing of cache memories," International Test Conference 1995, pp. 384-93.

[131] B. Nadeau-Dostie, A. Silburt, V.K. Agarwal, "Serial interfacing for embedded-memory testing," IEEE Design & Test of Computers, 4/1990, pp. 52-63.

[132] A.J. van de Goor, "Automatic computation of test length for pseudo-random memory tests," IEEE International Workshop of Memory Technology 1995, pp. 56-61.

[133] A. Krasniewski, K. Gaj, "Is there any future for deterministic self test of embedded RAMs," IEEE Computer Society Press, 4/1993, pp. 159-68.

[134] Y. Zorian, "BIST for embedded memories," EE-Evaluation Engineering, 9/1995, pp. 122-3.

[135] W.K. Al-Assadi, A.P. Jayasumana, Y.K. Malaiya, "On fault modeling and testing of content-addressable memories," IEEE Memory Technology, Design, and Test Workshop 1994, pp. 78-83.

[136] K. Lin, C. Wu, "Functional testing of content-addressable memories," IEEE Memory Technology, Design, and Test Workshop 1998, pp. 70-5.

[137] P.R. Sidorowicz, "Modeling and testing transistor faults in content-addressable memories," IEEE Memory Technology, Design, and Test Workshop 1999, pp. 83-90.

[138] R.D. Adams, P. Shephard III, "Silicon on insulator technology impacts on SRAM testing," VLSI Test Symposium 2000, pp. 43-47.

[139] Y. Wu, L. Calvin, "Shadow write and read for at-speed BIST of TDM SRAMs," International Test Conference 2001, pp. 985-94.

[140] K. Zarrineh, R.D. Adams, et al., "Self test architecture for testing complex memory structures," International Test Conference 2000, pp. 547-56.

[141] P. Mazumder, K. Chakraborty, *Testing and Testable Design of High-Density Random-Access Memories*, Kluwer, 1996.

[142] S. Hamdioui, "Testing multi-port memories: Theory and practice," Ph.D. Dissertation, Delft University, 2001.

[143] J. Zhao, et al., "Detection of inter-port faults in multi-port static RAMs," VLSI Test Symposium 2000, pp. 297-302.

[144] M.L. Bushnell, V.D. Agrawal, *Essentials of Electronic Testing: For Digital, Memory & Mixed Signal VLSI Circuits*, Kluwer, 2000.

[145] S.K. Jain, C.E. Stroud, "Built-in self-testing of embedded memories," IEEE Design & Test, 10/86, pp. 27-37.

[146] Y. Zorian, "BIST for embedded memories," EE-Evaluation engineering, 9/1995, pp. 122-3.

[147] C.E. Stroud, *A Designer's Guide to Built-In Self-Test*, Kluwer, 2002.

[148] B.F. Cockburn, Y.F. Nicole Sat, "Transparent built-in self-test scheme for detecting single V-coupling faults in RAMs," IEEE Memory Technology, Design, and Test Workshop 1994, pp. 119-24.

[149] K. Thaller, "A highly-efficient transparent online memory test," International Test Conference 2001, pp. 230-9.

[150] H. Bonges III, R.D. Adams, et al., "A 576K 3.5ns access BiCMOS ECL static RAM with array built-in self-test," IEEE Journal of Solid-State Circuits, Vol. 27, No. 4, 4/1992, pp. 649-56.

[151] R. Mookerjee, "Segmentation: a technique for adapting high-performance logic ATE to test high-density, high-speed SRAMs," IEEE Workshop on Memory Test 1993, pp. 120-4.

[152] P.H. Bardell, W.H. McAnney, J. Savir, *Built-In Test for VLSI*, Wiley, 1987.

[153] R. David, A. Fuentes, B. Courtois, "Random pattern testing versus deterministic testing of RAMs," IEEE Trans. on Computers, Vol. 38, No. 5, 5/89, pp. 637-50.

[154] W.W. Peterseon, E.J. Weldon, *Error-Correcting Codes*, MIT Press, 1972.

[155] A.J. van de Goor, "Automatic computation of test length for pseudo-random memory tests," IEEE International Workshop on Memory Technology 1995, pp. 56-61.

[156] V.N. Yarmolik, M. Nicolaidis, O. Kebichi, "Aliasing free signature analysis for RAM BIST," International Test Conference 1994, pp. 368-77.

[157] M. Franklin, K.K. Saluja, K. Kinoshita, "Design of a RAM with row/column pattern sensitive fault detection capability," International Test Conference 1989, pp. 327-36.

[158] J. van Sas, et al., "BIST for embedded static RAMs with coverage calculation," International Test Conference 1993, pp. 339-47.

[159] X. Li, F. Yang, "A cost-effective BIST scheme for embedded RAMs," International Conference on CAD and Computer Graphics 1993, Vol. 2, pp. 647-51.

[160] R.D. Adams, et al., "A 370-MHz memory built-in self-test state machine," European Design and Test Conference 1995, pp. 139-41.

[161] C. Hunter, et al., "The PowerPC 603 microprocessor: an array built-in self test mechanism," International Test Conference 1994, pp. 388-94.

[162] P. Camurati, et.al., "Industrial BIST of embedded RAMs," IEEE Design & Test of Computers, Fall 95, pp. 86-95.

[163] M. Franklin, K.K. Saluja, "Embedded RAM testing," IEEE Memory Technology, Design, and Test Workshop 1995, pp. 29-33.

[164] R. Dekker, F. Beenker, A. Thijssen, "Fault modeling and test algorithm development for static random access memories," International Test Conference 1988, pp. 353-61.

[165] V.D. Agrawal, C.R. Kime, K.K. Saluja, "A tutorial on built-in self-test," IEEE Design & Test of Computers, 6/1993, pp. 69-77.

[166] Thomas J. Eckenrode, personal communication July 2002.

[167] Garrett S. Koch, personal communication July 2002.

[168] R.D. Adams, R. Mazzarese, "A high-speed pipelined memory built-in self-test state machine," Dartmouth – Thayer School of Engineering HDL-Based System Design Project, May 1996.

[169] M. Nicolaidis, O. Kebichi, V. Castor Alves, "Trade-offs in scan path and BIST implementations for RAMs," Journal of Electronic Testing: Theory and Applications, Vol. 5, No. 4, 5-6/1994, pp. 147-57.

[170] R. Gibbins, R.D. Adams et al., "Design and test of a 9-port SRAM for a 100Gb/s STS-1 switch," IEEE Memory Technology, Design, and Test Workshop 2002.

[171] V. Castro Alves, et al., "Built-in self-test for multi-port RAMs," International Conference on Computer Aided Design 1991, pp. 248-51.

[172] Y. Wu, S. Gupta, "Built-in self-test for multi-port RAMs," Asian Test Symposium 1997, pp. 398-403.

[173] J. Zhu, "An SRAM built-in self-test approach," EE-Evaluation engineering, 8/1992, pp. 90-3.

[174] H. Koike, T. Takeshima, M. Takada, "A BIST scheme using microprogram ROM for large capacity memories," International Test Conference 1990, pp. 815-22.

[175] M.H. Tehranipour, Z. Navabi, S.M. Fakhraie, "An efficient BIST method for testing of embedded SRAMs," International Symposium on Circuits and Systems 2001, pp. 73-6.

[176] K. Zarrineh, R.D. Adams et al., "Self test architecture for testing complex memory structures," International Test Conference 2000, pp. 547-556.

[177] J. Dreibelbis, et al., "Processor-based built-in self-test for embedded DRAM," IEEE Journal of Solid-State Circuits, Vol. 33, No. 11, 11/1998, pp. 1731-40.

[178] A. Benso, et al., "A programmable BIST architecture of clusters of multiple-port SRAMs," International Test Conference 2000, pp. 557-66.

[179] D. Youn, T. Kim, S. Park, "A microcode-based memory BIST implementing modified march algorithm," Asian Test Symposium 2001, pp. 391-5.

[180] J. Barth, et al., "Embedded DRAM design and architecture for the IBM 0.11 um ASIC offering," IBM Journal of Research and Development, to be published.

[181] P. Jakobsen, et al., "Embedded DRAM built in self test and methodology for test insertion," International Test Conference 2001, pp. 975-84.

[182] R. Gibbins, R.D. Adams et al., "Design and Test of a 9-port SRAM for a 100Gb/s STS-1 Switch," IEEE Memory Technology, Design, and Test Workshop 2002, pp. 83-7.

[183] IEEE Memory Technology, Design, & Test Workshop 2001, Panel on Memory Redundancy & Repair Challenges.

[184] E. Rondey, Y. Tellier, S. Borri, "A silicon-based yield gain evaluation methodolgy for embedded-SRMAs with different redundancy scenarios," IEEE Memory Technology, Design, and Test Workshop 2002, pp. 57-61.

[185] H. Kikukawa, et al., "0.13-um 32-Mb/64-Mb embedded DRAM core with high efficient redundancy and enhanced testability," IEEE Journal of Solid-State Circuits, Vol. 37, No. 7, 7/2002, pp. 932-40.

[186] T. Namekawa, et al., "Dynamically shift-switched dataline redundancy suitable for DRAM macro with wide data bus," IEEE Journal of Solid-State Circuits, Vol. 35, No. 5, 5/2000, pp. 705-12.

[187] J. Barth, et al., "A 300MHz multi-banked eDRAM macro featuring GND sense, bit-line twisting and direct reference cell write," International Solid State Circuits Conference 2002, pp. 156-7.

[188] M. Yamaoka, et al., "A system LSI memory redundancy technique using an ie-Flash (inverse-gate-electrode flash) programming circuit," IEEE Journal of Solid-State Circuits, Vol. 37, No. 5, 5/2002, pp. 599-604.

[189] M. Ouellette, et al., "On-chip repair and an ATE independent fusing methodology," International Test Conference 2002.

[190] J. Wee, et al., "A post-package bit-repair scheme using static latches with bipolar-voltage programmable anti-fuse circuit for high-density DRAMs," IEEE Journal of Solid-State Circuits, Vol. 37, No. 2, 2/2002, pp. 251-4.

[191] T. Takahashi, et al., "A multigigabit DRAM technology with $6F^2$ open-bitline cell, distributed overdriven sensing and stacked-flash fuse," IEEE Journal of Solid-State Circuits, Vol. 36, No. 11, pp. 1721-7.

[192] K. Chakraborty, "A physical design tool for built-in self-repairable RAMs," IEEE Transactions on Very Large Scale Integration (VLSI) Systems, Vol. 9, No. 2, 4/2001, pp. 352-64.

[193] S. Tanoi, "On-wafer BIST of a 200Gb/s failed-bit search for 1Gb DRAM," International Solid State Circuits Conference 1997, pp. 70-1.

[194] Y. Wu, S. Gupta, "Built-in self-test for multi-port RAMs," Asian Test Symposium 1997, pp. 398-403.

[195] Y. Nagura, "Test cost reduction by at-speed BISR for embedded DRAMs," International Test Conference 2001, pp. 182-7.

[196] D.K. Bhavsar, "An algorithm for row-column self-repair of RAMs and its implementation in the Alpha 21264," International Test Conference 1999, pp. 311-8.

[197] A. Meixner, J. Banik, "Weak write test mode: An SRAM cell stability design for test technique," International Test Conference 1996, pp. 309-18.

[198] H. Pilo, R.D. Adams, et al., "Bitline contacts in high density SRAMs: Design for testability and stressability," International Test Conference 2001, pp. 776-82.

[199] J. Brauch, J. Fleischman, "Design of cache test hardware on the HP PA8500," International Test Conference 1997, pp. 286-93.

[200] Y. Wu, L. Calin, "Shadow write and read for at-speed BIST of TDM SRAMs," International Test Conference 2001, pp. 985-94.

[201] R. Gibbins, R.D. Adams et al., "Design and Test of a 9-port SRAM for a 100Gb/s STS-1 Switch," IEEE Memory Technology, Design, and Test Workshop 2002, pp. 83-7.

[202] H. Pilo, et al., "Design-for-test methods for stand-alone SRAMs at 1Gb/s/pin and beyond," International Test Conference 2000, pp. 436-43.

[203] K. Takeda, et al., "Quasi-worst-condition built-in-self-test scheme for 4-Mb loadless CMOS four-transistor SRAM macro," Symposium on VLSI Circuits 2001, pp. 229-30.

[204] J. Yamada, et al., "A 128-kb FeRAM macro for contact/contactless smart-card microcontrollers," IEEE Journal of Solid-State Circuits, Vol. 37, No. 8, 8/2002, pp. 1073-9.

[205] H. Kikukawa, et al., "0.13-um 32-Mb/64-Mb embedded DRAM core with high efficient redundancy and enhanced testability," IEEE Journal of Solid-State Circuits, Vol. 37, No. 7, 7/2002, pp. 932-40.

[206] K. Zarrineh, S.J. Upadhyaya, V. Chickermane, "System-on-chip testability using LSSD scan structures," IEEE Design & Test of Computers, 5-6/2001, pp. 83-97.

[207] A.J. van de Goor, G.N. Gaydadjiev, "An analysis of (linked) address decoder faults," IEEE Memory Technology, Design, and Test Workshop 1997, pp. 13-9.

[208] D. Niggemeyer, E.M. Rudnick, "Automatic generation of diagnostic march tests," VLSI Test Symposium 2001, pp. 299-304.

[209] http://www.tuxedo.org/~esr/jargon/html/entry/idempotent.html

[210] V. Castro Alves, et al., "Built-in self-test for multi-port RAMs," International Conference on Computer Aided Design, 11/1991, pp. 248-51.

[211] B.F. Cockburn, Y.F. Nicole Sat, "Transparent built-in self-test scheme for detecting single V-coupling faults in RAMs," IEEE International Workshop on Memory Technology, Design, and Testing 1994, pp. 119-24.

[212] J.P. Hayes, "Detection of pattern-sensitive faults in random-access memories," IEEE Trans. on Computers, Vol. C-24, No. 2, 1975, pp.150-7.

[213] J.P. Hayes, "Testing memories for single-cell pattern-sensitive faults in semiconductor random-access memories," IEEE Trans. on Computers, Vol. C-29, No. 3, 1980, pp. 249-54.

[214] M.G. Karpovsky, V.N. Yarmolik, "Transparent memory testing for pattern sensitive faults," International Test Conference 1994, pp. 860-869.

[215] D.S. Suk, S.M. Reddy, "Test procedures for a class of pattern-sensitive faults in semiconductor random-access memories," IEEE Trans. on Computers, Vol. C-29, No. 6, pp. 419-29.

[216] K.K. Saluja, K. Kinoshita, "Test pattern generation for API faults in RAM," IEEE Trans. on Computers, Vol. C-34, No. 3, 1985, pp. 284-7.

[217] P. Camurati, et al., "Industrial BIST of embedded RAMs," IEEE Design & Test of Computers, Fall 1995, pp. 86-95.

[218] M.S. Abadir, H.K. Reghbati, "Functional testing of semiconductor random access memories," Computing Surveys, Vol. 15, No. 3, 9/83, pp. 175-98.

[219] B. Nadeau-Dostie, A. Silburt, V.K. Agarwal, "Serial interfacing for embedded-memory testing," IEEE Design & Test of Computers, 4/1990, pp. 52-63.

[220] S. Griep, et al., "Application of defect simulation as a tool for more efficient failure analysis," Quality and Reliability Engineering Int., Vol. 10, 1994, pp. 297-302.

[221] Y. Lai, "Test and diagnosis of microprocessor memory arrays using functional patterns," M.S. Thesis, MIT, May 1996.

[222] S. Hamdioui, A. J. van de Goor, et al., "Realistic fault models and test procedure for multi-port SRAMs," IEEE Memory Technology, Design, and Test Workshop 2001, pp. 65-72.

[223] S. Hamdioui, A. J. van de Goor, et al., "Detecting unique faults in multi-port SRAMs," Asian Test Symposium 2001, pp. 37-42.

[224] J. Knaizuk, C.R.P. Hartmann, "An optimal algorithm for testing stuck-at faults in random access memories," IEEE Trans. on Computers, Vol. C-26, No. 11, 1977, pp. 1141-1144.

[225] R. Nair, "Comments on 'An optimal algorithm for testing stuck-at faults in random access memories,'" IEEE Trans. on Computers, Vol. C-28, No. 3, 1979, pp. 258-61.

[226] M.S. Abadir, J.K. Reghbati, "Functional testing of semiconductor random access memories," ACM Computing Surveys, Vol. 15, No. 3, 1983, pp. 175-98.

[227] A.J. van de Goor, *Testing Semiconductor Memories: Theory and Practice*, Wiley, 1991.

[228] M. Nicolaidis, O. Kebichi, V. Castor Alves, "Trade-offs in scan path and BIST implementations for RAMs," JETTA, Vol. 5, 1994, pp. 273-83.

[229] M. Marinescu, "Simple and efficient algorithms for functional RAM testing," IEEE Test Conference 1982, pp. 236-9.

[230] K. Zarrineh, S.J. Upadhyaya, "A new framework for automatic generation, insertion and verification of memory built-in self-test units," VLSI Test Symposium 1999, pp. 391-6.

[231] A.J. van de Goor, "March LA: A test for linked memory faults," European Design & Test Conference 1997, p. 627.

[232] S. Hamdioui, "Testing multi-port memories: Theory and practice," Ph.D. Dissertation, Delft University, 2001.

[233] R. Dekker, F. Beenker, A. Thijssen, "Fault modeling and test algorithm development for static random access memories," International Test Conference 1988.

[234] R. Dekker, F. Beenker, LFSR. Thijssen, "A realistic self-test machine for static random access memories," International Test Conference 1988, pp. 353-61.

[235] J. Zhu, "An SRAM built-in self-test approach," EE-Evaluation Engineering, 8/1992, pp. 90-3.

[236] C. Hunter, et al., "The PowerPC 603 microprocessor: an array built-in self test mechanism," International Test Conference 1994, pp. 388-94.

[237] J.H. de Jonge, A.J. Smeulders, "Moving inversions test pattern is thorough, yet speed," Computer Design, May 1976, pp. 169-73.

[238] R.D. Adams, R. Mazzarese, "A high-speed pipelined memory built-in self-test state machine," Dartmouth – Thayer School of Engineering HDL-Based System Design Project, May 1996.

GLOSSARY / ACRONYMS

ABIST – Array built-in self-test. This term is synonymous with memory built-in self-test.
AF – Address decoder fault.
Aggressor – Cell which causes erroneous operation in a victim cell.
ATE – Automated test equipment.
ATPG – Automatic test pattern generation.

Beta ratio – Typically the ratio between the pull-down strength and transfer strength in an SRAM cell.
BIST – Built-in self-test
Bit oriented – A memory which is accessed one bit at a time.
Bridging defect – A short between to signal lines.

CAM – Content addressable memory
CF – Coupling fault.

DRAM – Dynamic random access memory.

EEPROM – Electrically erasable programmable read only memory.

FeRAM – Ferroelectric random access memory.
Flash – A type of EEPROM where large portions of memory can be erased simultaneously.

Galloping pattern – Pattern which ping-pong addresses through all possible address transitions.

LFSR – Linear feedback shift register. Used to generate pseudo-random patterns.

Marching pattern – Sequentially addresses memory, leaving new data type in its wake.
MISR – Multiple input signature register.
MRAM – Magnetoresistive random access memory.

NPSF – Neighborhood pattern sensitive fault.

PROM – Programmable read only memory.

ROM – Read only memory.

SOC – System on chip.
SOI – Silicon on insulator.
SRAM – Static random access memory.

Ternary CAM – CAM which stores "1", "0", or "don't care" on a per bit basis.
TF – Transition fault.

Victim – Cell which is impacted by operation in aggressor cell.

Walking pattern – Sequentially addresses memory, leaving old data type in its wake.
Word oriented – A memory which is accessed one word at a time.

INDEX

ABOUT THE AUTHOR

R. Dean Adams has over 20 years of experience in the design and development of embedded and stand-alone memories, high-performance custom logic, and DFT/BIST circuits for IBM. He is currently a member of IBM's Design-For-Test (DFT) department, where he consults on numerous memory and advanced microprocessor design projects. Dean is a leading expert in high-performance memory BIST, as well as the modeling and testing of standard and custom logic for leading edge semiconductor technologies.

He holds Ph.D., M.S., and Master of Engineering Management degrees from Dartmouth, at the Thayer School of Engineering. He also has a B.S.E.E. degree from the University of Rhode Island. Dean is the inventor or co-inventor of over 20 patents, and has authored or co-authored numerous technical papers. Most of these papers and patents relate to memory testing, memory self-test, and memory design.

CPSIA information can be obtained
at www.ICGtesting.com
Printed in the USA
LVHW021321260720
661559LV00019B/991

9 781402 072550